THE
INVISIBLE UNIVERSE

Dark Matter, Dark Energy, and the
Origin and End of the Universe

THE INVISIBLE UNIVERSE

Dark Matter, Dark Energy, and the Origin and End of the Universe

Antonino Del Popolo
Università degli Studi di Catania, Italy

 World Scientific

NEW JERSEY • LONDON • SINGAPORE • BEIJING • SHANGHAI • HONG KONG • TAIPEI • CHENNAI • TOKYO

Published by

World Scientific Publishing Co. Pte. Ltd.
5 Toh Tuck Link, Singapore 596224
USA office: 27 Warren Street, Suite 401-402, Hackensack, NJ 07601
UK office: 57 Shelton Street, Covent Garden, London WC2H 9HE

British Library Cataloguing-in-Publication Data
A catalogue record for this book is available from the British Library.

First published 2021 (hardcover)
Reprinted 2022 (in paperback edition)
ISBN 978-981-125-263-1 (pbk)

THE INVISIBLE UNIVERSE
Dark Matter, Dark Energy, and the Origin and End of the Universe

Copyright © 2021 by World Scientific Publishing Co. Pte. Ltd.

All rights reserved. This book, or parts thereof, may not be reproduced in any form or by any means, electronic or mechanical, including photocopying, recording or any information storage and retrieval system now known or to be invented, without written permission from the publisher.

For photocopying of material in this volume, please pay a copying fee through the Copyright Clearance Center, Inc., 222 Rosewood Drive, Danvers, MA 01923, USA. In this case permission to photocopy is not required from the publisher.

ISBN 978-981-122-943-5 (hardcover)
ISBN 978-981-122-944-2 (ebook for institutions)
ISBN 978-981-122-945-9 (ebook for individuals)

For any available supplementary material, please visit
https://www.worldscientific.com/worldscibooks/10.1142/12074#t=suppl

Desk Editor: Rhaimie Wahap

Typeset by Stallion Press
Email: enquiries@stallionpress.com

CONTENTS

Introduction   vii

Chapter 1	A Brief History of the Big Bang Theory	1
Chapter 2	The Primordial Universe	29
Chapter 3	How do We Know that Dark Matter Exists?	63
Chapter 4	The Harmony of the World	87
Chapter 5	What Is Dark Matter?	99
Chapter 6	The Golden Age of Particle Physics: The Standard Model	115
Chapter 7	Dark Matter Candidates	153
Chapter 8	Detection of Dark Matter	175
Chapter 9	Dark Energy	195
Chapter 10	End of the Universe	223

Appendices
Appendix A: Inflation   243
Appendix B: Formation of Structures   249
Appendix C: Higgs Mechanism in More Detail   254
Appendix D: The Mass of Neutrinos   261
Bibliography   263
Index   267

INTRODUCTION

Everyone is a moon, and has a dark side which he never shows to anybody.

— Mark Twain

Humans are an inseparable part of the universe, material that emerged from the Big Bang, whose bowels, muscles, lungs, and heart are associated with the elements forged in the nuclei of the stars. It is therefore no coincidence that the sky has intrigued man from prehistory, like the cave paintings, representing gods and heroes, as well as temples and ancient astronomical observatories testify. This is a peculiarity of our species, classified by Linnaeus as *Homo sapiens*. The Greeks instead indicated the human being with the term *anthropos*, which in an etymological interpretation would mean "the one who looks up."[1]

Of all the species that lived and live on this planet, as far as we know, man is the only being capable of consciousness, of observing the sky with awareness, understanding its beauty, and asking questions on the meaning of life and on that of death, on the origin and on the end, both personal and that of the universe. Moreover, the term *homo* (man) has the same root as the term *humus*, earth, mud. So the *Homo anthropos* is a sort of bridge between earth and sky, endowed with an

---

[1] In fact the term *anthropos* could be split into *anò* (up), *athreo* (I look), and *ops* (eye). Interpretation confirmed by Ovid's *Metamorphoses*, in which we read, "Os homini sublime dedit, coelumque tueri iussit [deus]" (Man had as a gift a face turned upwards and his gaze aims at the sky and rises towards the stars). Book I, vv. 85–86.

innate curiosity that leads him to try to study and understand the mysteries of nature. The immense number of human beings who have lived to date on Earth are united by heaven. Languages and civilizations have changed or disappeared, together with the religions that have addressed the way of seeing life and its purposes. Today's social organization is certainly very different from that of some tens of thousands of years ago. Evolution in all areas of life has changed and eliminated what were firm points in the way of thinking. In all these changes, a constant is the immutability of the sky, which has remained the same, except for minor changes. The Moon always has the same phases, and the Sun follows its annual cycle linked to the seasonal changes. The position of the constellations, visible in different periods of the year, is still almost unchanged. Ultimately, human beings from 30,000 years ago lived under our own sky, saw the same constellations, and the same stars in similar positions.

Obviously, although the sky is the same, it is perceived differently: different is the understanding of the objects and phenomena of which it is constituted, together with their meaning. For the Australian Aborigines, the Sun was a woman who woke up in her camp in the east and lighted a fire, then lighted a torch that she carried around the sky. Before starting the journey, she decorated herself with red ocher, which, dispersing in the air, dyed the cloud with red. That was the dawn. Once she reached the west, she changed her makeup, dying the clouds with yellow and red, and that was the sunset. Finally, the Sun-Woman made an underground journey to return to her camp. For the Indo-European peoples, the Sun and the Moon moved in the sky on horse-drawn carriages, guided by a charioteer. The Vikings explained the eclipses of the Sun by referring to a wolf, Skoll, who ran after the Sun god, Sol, and, once captured, tore him apart. When this happened, people made a great noise with pots and pans to frighten the wolf to make the Sun return. The same thing happened during the lunar eclipse: the wolf Hati devoured the Moon (Mani). The Persians believed that eclipses were divine punishments, and for the Romans it was unthinkable that eclipses were due to natural causes. The understanding of the phenomenon and the prediction of eclipses is attributed to the Greeks (Thales 585 BC), although it seems that the ancient Chaldeans, 2500 years ago, already knew the 18-year cycle called the *Saros cycle*, at the end of which eclipses repeat.

Skoll and Hati (2009). Credit: Akreon. Painting.

Ignorance of natural events led men to create myths and invent gods who supervised every aspect of life. Although several Greek and Magna Graecia philosophers (present-day Sicily) like Pythagoras, Archimedes, Anaximander, Empedocles, and Aristarchus, started to have a more scientific view of the world, it was often not accepted for cultural or political reasons. Epicurus did not accept atomism. Even Kepler, many centuries later, believed that planets were sentient beings. Descartes and Newton, although they believed that natural laws had been established by God and that the universe regulates itself by following these laws, believed that God could change them at any time. For Newton, God continually intervened to make the planets follow the right orbits. With the evolution of research and scientific thought, the epexegetic role of divinities and myths has been increasingly reduced. The roles of religion and science have clearly separated.

As we said, the sky unites us and our ancestors. Until a century ago, the universe was considered to be constituted just by stars, planets, and comets.

Today, the universe is much more complex for us than for our ancestors: black holes, neutron stars, white dwarfs, supernovae, various types of nebulae, galaxies, clusters, superclusters of galaxies, and much more. The most shocking fact is that today we know that all the innumerable objects we observe in our universe represent only about 5% of the matter that constitutes it, the "visible" one. Our advantage in knowledge compared to our ancestors sees a significant reduction if we think that we have a limited knowledge even of the small part of the universe we see. We also don't even know what it's actually made of. We can only quantify our degree of ignorance about the universe with a percentage, 95%, and this ignorance has a name: dark matter and dark energy. Of the first we have some ideas: these are probably elusive particles that weakly interact with matter. Of the second, which started to manifest itself just a few billion years ago, we know even less; it is possible that it is energy of the vacuum that acts as repulsive gravity, but in fact we are groping in the dark.

Cosmology studies the universe in its entirety, trying to understand its origin, its structure, and its evolution.

In this book, I will describe the cosmology of the so-called Big Bang, and an alternative to it, which explains how the universe was born and evolved, going from a state where energy dominated in a hot environment to the crystallization of this energy in the galaxies and structures that we observe. We will see how the universe originated from an accelerated expansion phase, known as *inflation* (see section 2.2, and Appendix A), which made the universe grow in unimaginably small times, from microscopic to macroscopic dimensions, smoothing out its geometry and transforming the energy of the vacuum into matter and energy.

We will see how the study of galaxies, clusters of galaxies, the deviation of the light produced by the masses, and the *microwave background radiation* that pervades the whole universe helped us to understand that much of the matter that constitutes the universe is invisible, and we will discuss several assumptions about the nature of that matter.

Dark matter is made up of particles, not similar to those that make up ordinary matter, new particles foreseen by theories that go beyond the *standard model of elementary particles*, produced, revealed, and studied in the largest accelerating machines such as LHC (Large Hadron Collider) at CERN (European Center for Nuclear Research) in Geneva. We will discuss this model, the history of its construction, the basic ideas on which it is built, and how the world we see, apparently so solid, is actually made up of fields and waves and how even the elementary forces of nature are born from fields and "symmetries." We will see that the world is very different from how we perceive it. We will then venture into the "zoo" of dark matter particles candidates and discuss their characteristics. We will talk about particles whose existence is dictated by a particular symmetry, the *supersymmetry*, and we will venture into the byways of extra dimensions in search of dark matter particles, describing the methods by which the hunt for this modern Grail is carried forward.

The curtain will then open on the mystery of dark energy and on the hypotheses formulated on its nature to conclude with the role it plays in the evolution and end of the universe. Energy, for the properties currently known, would lead the universe to thermal death, a universe in which only photons move in huge and cold spaces. If dark energy revealed different characteristics in the future, other scenarios could open up, in which the universe could recollapse into a *Big Crunch* or have a cyclic nature that, after birth, evolution, and end, would see it reborn like the phoenix from its ashes.

Chapter 1

A BRIEF HISTORY OF THE BIG BANG THEORY

The history of astronomy is the history of receding horizons.

— Edwin Hubble

1.1 A Static Universe

The conception of the cosmos from ancient times to the Middle Ages and beyond was based on Aristotelian physics and the astronomical models of Ptolemy. This universe was static; nothing changed: the planets and the stars moved following unchanging and eternal cycles. The ideas of Copernicus and Giordano Bruno and the publication of *Principia Matematica* by Newton in 1687 changed the idea of the cosmos considerably, but not the idea of its static nature.

An unchangeable universe has a remarkable charm. A feature of Western thought was the preference for a worldview in which the universe was not affected by what was happening on Earth. Examples of this way of seeing things are the ancient *myth of the eternal return*, taken up in modern times even by Nietzsche, or the clockwork universe typical of Newtonian mechanics. The Newtonian view is reductionist in the sense that the explanation of the functioning of the cosmos is reduced to the application of physical laws. A generalized interpretation of reductionism results in the static nature of the universe. In fact, even if the planets move, they do so in accordance with fixed and eternal laws that, by repeating themselves, give rise to a universe always equal to itself. However, there is a problem: staticity conflicts with Newtonian mechanics. Gravity being attractive, as Richard Bentley pointed out in the late seventeenth century, collapse is the ultimate fate of all static and finite systems. It was not difficult for Newton to resolve the paradox. It was enough to assume that the universe was infinite and that the masses were evenly distributed, leading, for reasons of symmetry, to the cancellation of gravitational forces. Thus, Newton's universe was infinite, uniform, static and in unstable equilibrium. A second paradox, which would further undermine the Newtonian vision, was, however, lurking: the so-called Olbers paradox, according to which the fact that the night sky is not as clear as in the daytime is incompatible with a universe infinite and uniformly populated by stars.

The idea of a static universe was so hard to die that it was necessary to wait until the 20th century to see a change. Before Hubble's discovery of the expanding universe, the idea of a static universe was widely accepted by 20th-century scientists, including Einstein. Two years after the publication of his treatise on the theory of general relativity, in 1917, Einstein applied it to the universe and found results in contradiction with his prejudices and those of his time: namely, the static nature of the universe. He found that his equations predicted an expanding or contracting universe. Showing, in that case, little confidence in his equations, and following the generalized prejudice that the universe was static, he introduced into his equations a constant, the cosmological constant Λ, which, acting repulsively, counteracted the gravitational attraction and compelled the universe to remain static.

The problem that Einstein had, using his equations, was very similar to what Bentley had identified centuries earlier in the context of Newtonian mechanics, that is, any finite and static system collapses sooner or later. Einstein could not solve the problem in Newton's way because of the intrinsic structure of his model. Einstein's universe had a peculiar geometric structure, finite and without borders. In geometric terms it was a *3-sphere*, that is, a three-dimensional sphere incorporated in a four-dimensional space. To understand this better, the surface of a balloon is a two-dimensional sphere, called a

2-sphere, in a three-dimensional space. Dante Alighieri had already imagined this structure: his *Paradise* has the structure of a 3-sphere, as the mathematician Mark Peterson noticed in 1979.[1] His universe was finite but limitless, borderless like a sphere, and since gravity is attractive, alas, such a universe could not be static.

Therefore, while Newton had managed to solve the problem and make the universe static with the hypothesis that it was infinite, uniform and symmetrical, this could not have happened in Einstein's geometry. So, in order to resolve the contradiction, he introduced the cosmological constant that was tailored to block the tendency for expansion or contraction of his universe.

The result was a static and closed universe. Furthermore, the cosmological constant, Λ, was a sort of timid presence that made its effects known only on very large scales — cosmological scales. It seemed like a good solution to the problem. Einstein believed that in the case of a non-zero cosmological constant and zero-matter density, there were no solutions to his equations. This was not the case, and in the same year Willem de Sitter showed that a universe with a cosmological constant and devoid of matter would expand exponentially. For two years Einstein tried to find an error in de Sitter's solution but found none.

1.2 Friedman's Universe

Nature, however, is not easily harnessed and sooner or later tricks come to the light. So in 1922, a Russian meteorologist, physicist-mathematician by training, multifaceted character by nature, who worked in the observatory of today's St. Petersburg, published an article titled "On the Curvature of Space" in a German scientific journal, in times when physics was synonymous with Germany. The solutions he obtained were not different from those initially found by Einstein: they

---

[1] This intuition was later taken up in a 2006 book by Horia-Roman Patapievici, *The Eyes of Beatrice: What Dante's World Really Was Like*.

provided a homogeneous, isotropic and non-static universe. Unlike Einstein, Friedman was not frightened by the results. The solutions of Friedmann's equations, that is, the possible types of universe that the equations describe, are of three kinds, as shown in Fig. 1.1.

The three types of universes, that is, the solutions of Einstein's equations, depend on the density of mass-energy (remember that mass and energy are two sides of the same coin, as shown by Einstein with his famous formula $E = mc^2$) of the universe. It is easy to identify a particular density value called critical density, ρ_c, whose value is equal to 8.6×10^{-30} g/cm³, i.e. 0.0000000000000000000000000000086 g/cm³ (see Box 1), or more simply 5 hydrogen atoms per cubic meter. This density takes this name because it discriminates between the three different types of geometry in the universe. The three possible types of universe are shown in Fig. 1.1. Each curve represents the increase in distance between galaxies with time. In the vertical axis of Fig. 1.1 is plotted the so-called expansion parameter, which represents the relative expansion of the universe and can be considered as the average distance between galaxies. In the horizontal axis we have time.

If the average density of the universe, ρ, is greater than the critical density, ρ_c, the universe expands to a maximum

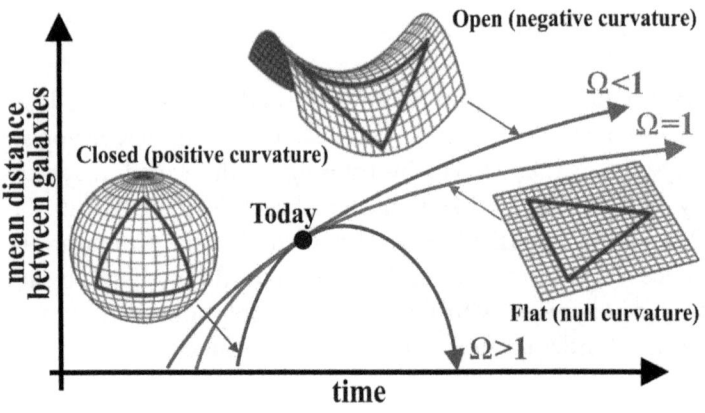

Fig. 1.1. The three Friedman universes.

BOX 1
POWERS OF TEN
In order to write large and small numbers in a simpler and more compact manner, scientists use the powers of 10 notation. A power of 10 is the number followed by a number of zeros equal to the exponent. For example, if we want to write 1000, that is, 1 followed by 3 zeros, we will write 10^3. The mass of the Sun is 2×10^{30} kg, that is, 2 followed by 30 zeros: 2000000000000000000000000000000.
In case of small numbers, for example a thousandth, $1/1000 = 0.001$, namely 1 preceded by 3 zeros, we write 10^{-3}.
In summary, 10^n stands for 1 followed by n zeros, while 10^{-n} stands for 1 preceded by n zeros.
MASS UNITS USED IN ASTROPHYSICS
The masses of celestial objects are generally large. The mass of the Sun in kilograms is 2×10^{30} and is often used as a unit of measurement for the masses and is written as 1 $M_\odot$.
Then: 1 $M_\odot = 2 \times 10^{30}$ kg
A star with a mass 8 times greater than that of the Sun will be indicated as 8 $M_\odot$.
DISTANCE UNITS IN ASTROPHYSICS
Similarly, to masses, distances in astrophysics are expressed in different units:
Astronomical unit, indicated with A.U., is the Earth–Sun distance, ~149 597 870 700 km
Light-Year, indicated with l.y., is the distance covered in one year by light: 9 460 730 472 581 km
Parsec, indicated with pc, is equal to 3.26 light-years, that is 3.26×10^{16} m. Parsec multiples are kiloparsec, (kpc, 1000 pc), megaparsec (Mpc, 10^6 pc), and gigaparsec (Gpc, 10^9 pc).
TEMPERATURE AND ENERGY UNITS
In this book, temperatures are measured in Kelvin, and indicated with K. To obtain the temperatures in degree Celsius, °C, one has just to subtract from the value of the temperature in Kelvin the quantity 273.16.
As a measure of energy, we will use electronvolt, eV, and its multiples: kiloelectronvolt (keV, 1000 eV), gigaelectronvolt (GeV, 10^6 eV), and teraelectronvolt (TeV, 10^9 eV).
The electronvolt can also be expressed in terms of the Kelvin temperature: 1 eV $\approx$ 11,600 K.
The mass of an object may be expressed in electronvolt and its multiples. For example, the mass of the proton $m_p = 1.67 \times 10^{-27}$ kg ≈ 938 MeV/c^2.

radius and then re-collapses, as seen in the red curve of Fig. 1.1. Friedman called this solution "the periodic world." Space is similar to a three-dimensional sphere (the *3-sphere*, of which we have spoken), and for this reason the spatial geometry of the universe is called *spherical*. This universe is called the *closed universe*,[2] and has positive curvature. Unlike Euclidean geometry, in this universe the sum of the internal angles of a triangle is greater than 180 degrees, the space is finite and two parallel straight lines converge. If the density is less than the critical one, as seen in the green curve of Fig. 1.1, the universe will expand forever, and for this reason we speak of an *open universe* with negative curvature. Its spatial geometry is called *hyperbolic*, because it resembles that of a horse saddle, and space is infinite. In hyperbolic geometry the sum of the internal angles of a triangle is less than 180 degrees and two parallel straight lines converge in one direction while diverging in the other. In the case that the density is equal to the critical one, as in the blue line of Fig. 1.1, the universe will expand forever, the geometry is *Euclidean*, in which the sum of the internal angles of a triangle is 180 degrees, and two parallel lines do not meet. The curvature of such a universe is zero. This type of universe is also called *flat universe*, its geometry being like that of a plane.

In cosmology, for practical reasons, instead of using the critical density, ρ_c, the ratio between density ρ and critical density ρ_c is used. This ratio is called the *density parameter* and is indicated with $\Omega = \rho/\rho_c$. The density parameter also expresses the mass quantity of a certain species in terms of the critical density. For example, as we will see, baryonic matter (that of which we are made) constitutes only 5% of the matter in the universe, and therefore in terms of ρ_c we will write $\Omega_b = 0.05$. The three types of geometry that we have seen before and that is shown in Fig. 1.1 can be expressed in terms of the density parameter Ω, thus: the flat universe, with density equal to the critical one ($\rho = \rho_c$) has Ω equal to 1 ($\Omega = 1$); the hyperbolic one has Ω less than 1 ($\Omega < 1$) and the closed one has Ω greater than 1 ($\Omega > 1$).

The three behaviors of Friedman's solutions can be explained more intuitively using Newtonian mechanics. If we

---

[2] Curvature indicates how far a curve or object deviates from being flat.

throw a ball upwards with a speed lower than 11.2 km/s (called *escape velocity*, the minimum speed for an object to leave the Earth and overcome its gravitational attraction), the ball will fall back. This is what happens in Friedman's closed universe ($\Omega > 1$). If the ball moves faster than the escape velocity, it will no longer return to Earth. This corresponds to the case of a hyperbolic universe ($\Omega < 1$). If the velocity is equal to the escape velocity, the ball again will never go back to Earth and we are in the case ($\Omega = 1$).

In addition to showing that Einstein's universe was unstable (expanding or collapsing), Friedman was the first to say that the universe would somehow have a beginning, and he also estimated its age to be 10–20 billion years. The universe would have been born from what mathematicians and physicists call a *gravitational singularity*, that is, a point where the curvature tends to an infinite value. Examples of gravitational singularities, in addition to that linked to the birth of the Universe, are *black holes* — which we have all heard of — which swallow everything that passes within a certain radius, even light! For Friedman, however, this singularity was just something mathematical, not physical.

As we will see below, only several years later, in 1927, Lemaître gave a physical meaning to this gravitational singularity: the explosion of the so-called *primeval atom*, today known as the Big Bang. This term was coined by Fred Hoyle in an ironic way because he was the proponent of a model of universe competing with the Big Bang model: the so-called *steady-state theory*.[3] During a BBC broadcast, trying to explain the theory he said: "These theories were based on the hypothesis that all the matter in the universe was created in one Big Bang in a particular time in the remote past."

---

[3] The steady-state theory is a cosmological theory proposed by Fred Hoyle, Herman Bondi and Thomas Gold and based on the *perfect cosmological principle*, that is, on the idea that the universe is homogeneous and isotropic in space and time. The Universe is expanding and always has the same properties at any time and in any position. The decrease in density due to expansion is remedied by the continuous creation of matter.

Einstein reacted to Friedman's article by publishing a short note in which he claimed that the Russian's calculations were incorrect. Friedman wrote a letter to Einstein elaborating its derivation. Only a year later, on the 'insistence of Yuri Krutkov, a colleague of Friedman, Einstein read the letter and realized he was wrong. He published another article admitting that Friedman's solutions were correct, but according to him they were "devoid of physical sense." However, the latter sentence was deleted just before publication.

A decade had to pass before Einstein truly realized how much Friedman was right, and in 1931 he eliminated the cosmological constant from his equations and, as reported by Gamow in 1956, he considered that having introduced Λ into his equations was "the biggest blunder of my life."

In addition to previous results, Friedman discussed the problem of determining the geometry of the universe in 1923 using the triangulation[4] of distant objects such as Andromeda. This idea, similar to an idea by Gauss, was applied only a few decades ago, by means of some space missions, using much larger scales than those designed by Gauss.

Friedman had a short (1888–1925) but multifaceted life. Already in secondary school he made himself noticed, together with his close friend, the great mathematician Yakov Tamarkin. He studied at the University of Petrograd, now St. Petersburg, where he also had the great physicist Ehrenfest as a teacher. He was a professor at the University of Perm and then obtained a position in the Geophysical Observatory of today's St. Petersburg where he soon became director. During the First World War, onboard bomber aircrafts he personally verified his ballistics theories and obtained the Cross of Saint George for military merits. He even made a hot-air balloon record in 1925, reaching 7400 m for experiments. He died the same year of typhoid fever.

Despite his important results, Friedman's works had no or little impact on the astronomical community, also because he disappeared shortly after the publication of his works.

---

[4] Triangulation is a technique that allows you to calculate an unknown distance using the geometric properties of triangles.

1.3 Lemaître and the Expansion of the Universe

Ironically, the extension of Friedman's results and the foundations of the Big Bang theory are due to a Jesuit presbyter: Georges Lemaître.

Lemaître was born in Charleroi in 1894. Trained as an engineer, he joined the Belgian army in the First World War. After the war, he obtained his doctorate in 1920, then entered the seminary, and in 1923 he was ordained a priest. He continued his studies in Cambridge with Eddington, who greatly appreciated his talents. Then he continued his studies at the Massachusetts Institute of Technology, where he obtained a second doctorate. In 1925, he found Friedman's solutions to the equations of general relativity. He learned that they had already been obtained by Friedman in his meeting with Einstein in the famous Solvay Congress of 1927.

In the same year Lemaître had published in French his famous article "A Homogeneous Universe of Constant Mass and Increasing Radius Which Justifies the Radial Velocity of the Extragalactic Nebulae"[5] in an obscure Belgian magazine, *Annales de la societé Scientifique de Bruxelles*. In this paper he considered the dynamic solutions of the equations of general relativity from a more physical point of view than Friedman's, and he assumed that the radius of the universe could change arbitrarily over time. Among the results of the work, there is the relationship between the speed of recession, that is, the departure, of the extragalactic nebulae, and the distance at which they are found, today known as Hubble's law. The equation obtained was $v = kr$, with $k = 0.68 \times 10^{-27}$ cm^{-1}, namely 0.00000000000000000000000000068 cm^{-1} (cf. Box 1, Powers of Ten). The numerical values in the relationship were obtained using the velocity estimates of the Slipher and Stronberg nebulae. Lemaître did not give much importance to the result obtained, but nonetheless pointed out that the shift towards the red (redshift) of the nebulae should not

---

[5] "Un Univers homogène de masse constante et de rayon croissant rendant compte de la vitesse radiale des nébuleuses extra-galactiques".

BOX 2
LIGHT AND ELECTROMAGNETIC SPECTRUM

In nature there are different types of waves, such as those that are observed on a pond when we throw a stone into it. Sound, for example, is a wave. When a body vibrates, it transmits the vibrations to the air, which carries them up to the ear. Light is a wave that is produced by the oscillation of the electric and magnetic fields and for this reason it is called electromagnetic wave. $\vec{E}$ indicates the electric field. Similarly, $\vec{B}$ is the magnetic field (see Box 5 for a description of what a field is). It propagates at a speed of about 300,000 km/s (Fig. B2.1).

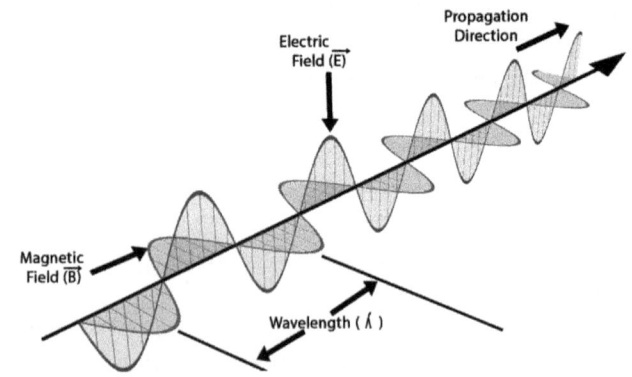

Fig. B2.1. The electomagnetic field.

All waves have a *wavelength*, the distance between two peaks, shown in Fig. B2.2.

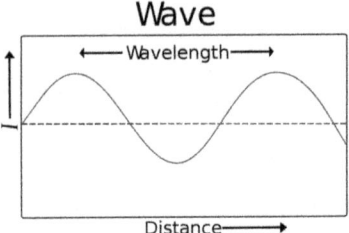

Fig. B2.2. A wave.

The *frequency*, ν, indicates the number of oscillations in a second and is linked to the *wavelength*, λ, by the relationship λν = wave velocity. A higher-frequency wave is more energetic than a lower-frequency one. Our eyes are sensitive to visible light, which has a wavelength between 390 and 700 nm, where nm stands for *nanometer*, or a billionth of a meter. The light we see is white, but it can be broken down, using a prism, into its components, from violet to red, as shown in

A Brief History of the Big Bang Theory

Fig. B2.3, which is called the continuous spectrum. Besides visible light, there are other electromagnetic radiation. The set of electromagnetic radiation constitutes the electromagnetic spectrum, which is made up of spectral regions, as shown in Fig. B2.3. From smaller to larger wavelengths, we have radio waves, microwaves, infrared, visible, ultraviolet, X-rays, and gamma rays.

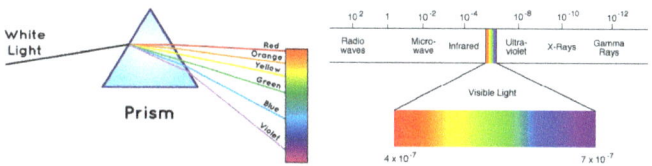

Fig. B2.3. light spectrum, and spectral bands.

Celestial objects emit in different regions of the spectrum depending on their temperature, chemical composition, and density. Electromagnetic radiation does not always behave like a wave. In some experiments as a wave in others as a particle. It also happens that particles behave like waves in certain experiments. In physics, this dual nature, wave and corpuscular, of the components of matter is referred to as *wave-particle dualism*. Light can therefore be considered as consisting of particles that move at the speed of light, without charge and mass called photons, usually indicated with the Greek letter γ.

be interpreted as a *doppler effect*,[6] that is, as the recession of galaxies in a static universe, but was due to the dynamics of the system itself. This is exactly today's point of view. The redshift observed in the spectra of the galaxies that move away from us is not due to the motion of the galaxies in space, but due to the fact that space expands and drags the galaxies with it.

The fact that the spectral lines (see Box 3, Absorption and Emission Spectrum) of stellar objects and galaxies was shifted towards red (see Box 3, Doppler Effect) was already known in the second half of the 19th century. Christian Doppler in 1842 and Armand Fizeau in 1845 had noticed that the spectral lines observed in the light of some stars were shifted towards the red region of the spectrum, and others towards the blue region.

This property was used by William Huggins to determine the velocities of the stars, while in 1912 Vesto Slipher applied

---

[6] Effect discovered by Christian Doppler in 1842 according to which the frequency of the waves emitted by a moving object are larger if the object approaches and vice versa (see Box 3, Doppler Effect).

BOX 3
ABSORPTION AND EMISSION SPECTRUM

In a simple, albeit incorrect, representation, a hydrogen atom is made up of a proton in the center and an electron rotating around it. There are different energy levels, which are indicated with the letter n. An electron can be found only on one energy level, that is, on $n = 1, 2, 3 \ldots$ Every atom has a unique number of electrons in a single configuration, that is, each element has its own distinct set of energy levels. This arrangement of energy levels constitutes the fingerprint of the atom. If a photon hits an atom and the energy of the photon is equal to the difference between two energy levels, then the electron on the lower level can absorb the photon and jump to the upper level, *exciting* the atom. If the energy is enough to free the electron, we speak of *ionization*. If the energy of the photon is not equal to the difference in energy between the two levels, it will not be absorbed. The same thing happens for the transition between a higher and a lower level. In the first case we will talk of photon absorption (Fig. B3.1).

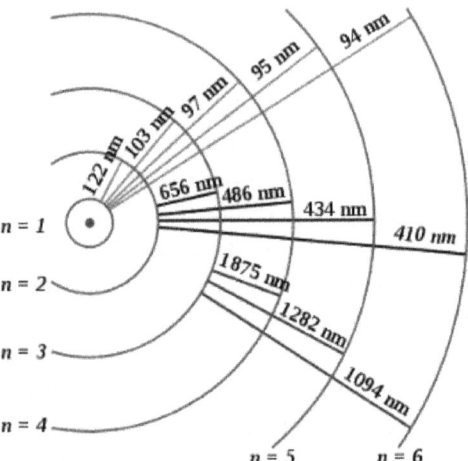

Fig. B3.1.   Energy levels of the hydrogen atom.

Let us now illuminate some gas whose light then pass through a prism (or a diffraction grating) (Fig. B3.2).

Fig. B3.2.

If we observe the light after passing it through the gas, we will observe gaps, dark empty areas. The dark lines correspond to energies (wavelengths) for which there is a difference in the energy levels of hydrogen. The absorbed photons appear as dark lines because the photons of these wavelengths have been absorbed (Fig. B3.2, right side).

Let's do another experiment, letting the light of a hot gas pass through a prism (or a diffraction grating) (Fig. B3.3). We will obtain an emission spectrum (Fig. B3.3, right side).

Fig. B3.3.

The emission spectrum will be the identical inverse of the absorption spectrum.

So the dark lines in the absorption spectrum correspond to the frequencies of light that have been absorbed by the gas. These dark lines correspond exactly to the frequencies of the emission spectrum.

In the experiment of absorption, we observe black lines. Each of them corresponds to different elements

In Fig. B3.4, some lines in the spectrum of the Sun are represented.

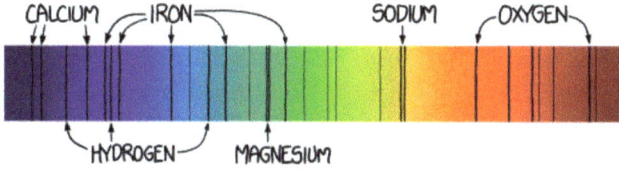

Fig. B3.4.

The lesson we learn from what has been seen is that each element has a typical spectrum, a bar code that identifies it without the possibility of making a mistake.

DOPPLER EFFECT

Galaxies are made up of a myriad of stars whose spectra combine into spectra similar to stellar ones. The study of the spectrum allows to establish, among other things, the velocity of the galaxies. The lines of the spectrum of an approaching galaxy move towards blue (*blueshift*), while those of a receding galaxy move towards red (*redshift*), as shown in Fig. B3.5.

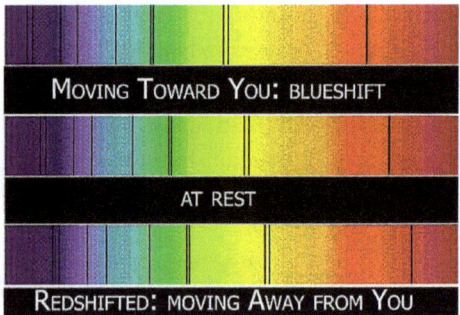

Fig. B3.5. Redshift.

Considering the wave as a ribbon, what happens is as if it was stretched when the object moves away and compressed when it approaches (Fig. B3.6).

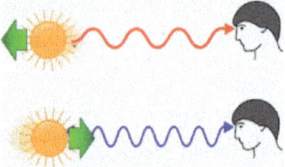

Fig. B3.6.

This effect is called the *Doppler effect* and takes its name from Christian Doppler, who discovered it in 1845. The Doppler effect applies to any wave, including sound. When an ambulance approaches with the sirens blaring, the sound (pitch) is more acute because, as the wave is compressed, the wavelength decreases and the frequency increases together with the energy, while when the ambulance moves away, the sound becomes less acute because of a contrary effect.

To test the effect, Doppler hired a group of musicians playing on a railroad car. Listening to the sound when the wagon approached and moved away, he verified the effect.

the method to the Andromeda galaxy by finding that it was moving towards us with a speed of 300 km/s. In 1917 he provided the results relating to the motion of 25 systems, of which only 4 approached us, and established that the average speed of the galaxies was 570 km/s.

Nonetheless, the ideas of the Lemaître's article were not accepted. Einstein, in a discussion with Lemaître, pointed out to him that from a mathematical point of view the article was irreproachable but from the point of view of physics it was abominable.

Lemaître's article, published in a little-known magazine, went unnoticed for three years, and gained notoriety only thanks to chance. At a meeting of the Royal Astronomical Society, de Sitter had discussed his latest results published in the conference proceedings. In the same conference, Eddington pointed out that an intermediate model between Einstein's and de Sitter's was needed, that is, a model that put a little movement in Einstein's model and a little matter in de Sitter's. It was then that Lemaître wrote to Eddington pointing out his work. Eddington advertised it by making it known to de Sitter and other colleagues and had it translated into English, making it acquire acclaim. The article became famous, but its fundamental part had been removed. In fact, in the translation of 1931 the data section and the discussion related to the Hubble law[7] had not been reported, since Lemaître, translator and editor of the article, did not consider that this point was important.

1.4 The Primordial Atom

Aside from Friedman, the idea of a universe that had a beginning was frowned upon by astronomers. For Friedman himself the singularity, represented only a "mathematical object." Only Eddington, in his 1928 book *The Nature of the Physical World*, mentioned a state from which the universe would originate and expand, although he was not convinced,

---

[7] The Hubble law is the proportionality relationship between the recession velocity of galaxies and their distance.

as he said that "the universe started with a bang."

Eddington's ideas were taken more seriously by Georges Lemaître, professor of astrophysics at the Catholic University of Leuven.

Lemaître in his 1931 article, "The Expanding Universe," stated that there could have been a static proto-universe in which "all the energy was in the form of electromagnetic radiation and suddenly condensed in matter." This idea was confirmed in the article "The Principle of the World from the Point of View of Quantum Theory," published in *Nature* in 1931, in which he reported:

> "We could conceive the beginning of the universe in the form of a unique atom, the atomic weight of which is the total mass of the universe. This highly unstable atom would divide in smaller and smaller atoms by a kind of super-radioactive process."

This basic idea was developed in a series of conferences held at the Belgian Astronomical Society, at the British Association for the Advancement of Science, and in the publication *The Expansion of Space*. In this article Lemaître wrote:

> "The initial expansion thus took place in three stages: a first period of rapid expansion in which the atom-universe was broken down into atomic stars; a period of slowing down, followed by a third period of accelerated expansion."

Lemaître's model showed that the universe had an origin but excluded the presence of a singularity. He did not discuss the origin of the primordial atom but assumed its existence before the so-called radioactive explosion. According to him, it made no sense to discuss the properties of the primordial atom before the moment of the explosion, since space and time would have started only after its disintegration. This is exactly the modern point of view: space and time originated with the Big Bang and therefore it makes no sense to discuss what existed before.

To quote St. Augustine, "the world was created with time, not in time," or as Stephen Hawking said, "wondering what happened before the Big Bang is like wondering what's north of the north pole."

While Lemaître's work on the expansion of the universe was well accepted, the primordial atom model was not welcomed in a positive way by the scientific community. To give just one example, in January 1933 Einstein attended a seminar by Lemaître on cosmic rays, erroneously considered by Lemaitre as a sort of "fossil radiation" that could give us information on the first moments of the universe. After the seminar Einstein discussed about cosmology with Lemaître and about his "primordial atom" he said that he did not like the idea because "it largely suggests the [theological] idea of creation." After a conference by Lemaître in Pasadena on the cosmology of the primordial atom, Einstein even went so far as to say, "This is the most beautiful and satisfying explanation of creation that I have ever heard." Einstein, like other scientists, made confusion between creation and beginning. For Lemaître the initial singularity was not the "creation" but the "natural beginning" of things. He always kept religion separate from science, so much so that in a personal meeting with Pope Pius XII he pointed out that the idea supported in 1951 by the Pope that science had demonstrated Genesis was wrong.

1.5 The Extragalactic Nature of the Nebulae

From an observational point of view, until the late 1920s the universe was thought to coincide with our galaxy. On April 26, 1920, Shapley and Curtis debated in Washington the thesis of the universe consisting only of our galaxy, defended by Shapley, and that ours was one of many galaxies, defended by Curtis. There were no loser and no winner.

The debate was resolved a few years later in three articles published in 1925, 1926, and 1929 by an American astronomer, Edwin Hubble. He showed the extragalactic character of NGC 6822, M33, and M31, giving reason to Curtis's thesis, which was also supported by other astronomers such as Öpik and

Lundmark. To achieve this he used special stars, the *Cepheids*, which we will discuss in a moment.[8]

A curious episode related to Shapley shows how scientists are human beings like everyone else.

When he was at Mount Wilson with Hubble and his collaborator, Humason, the latter presented to him a photographic plate showing a Cepheid in the Andromeda galaxy. Humason then marked the plate in order to find it later more easily. Shapley, having understood that the use of the Cepheid could show that the universe was much larger than his model, thus destroying its validity, deleted the mark, in an attempt to hide his mistakes about the existence of extragalactic objects. Obviously, Humason still knew the Cepheid's location and reported it to Hubble, who determined Andromeda's[9] distance.

Edwin Hubble was born in Marshfield in 1889. He studied mathematics and astronomy at the University of Chicago, and then obtained a master's degree in law from the Queen's College in Oxford. When he was back in the United States he started to teach physics, mathematics, and Spanish, which he had studied at Oxford, and only in 1914 decided to start his career as an astronomer. In 1917 he obtained his doctorate and in 1919 George Hale, director of the Mount Wilson Observatory in Pasadena, offered him a permanent position.

His knowledge in astronomy and physics was very limited. Nevertheless, he was able to obtain very important results in his career.

The results obtained by Hubble were made possible thanks to a fundamental discovery made by Henrietta Leavitt, who in 1908 published the article "1,777 Variables in the Magellanic Clouds" in the magazine *Annals of Harvard College Observatory* and "Periods of 25 Variable Stars in the Small Magellanic Cloud" in 1912, in the newsletter 173 of the observatory. In the articles, she showed the existence

---

[8] Hubble had started a search for exploding stars called Novae. Among the stars he saw one of these Novae but looking better he understood that it was a Cepheid that allowed him to determine the distance of Andromeda.

[9] Andromeda is also known as M31, the 31st galaxy in the Messier Catalog.

of a relationship between brightness and period of variable stars called Cepheids, whose name derives from δ Cephei, the second star of this type discovered. These stars pulsate periodically, with a period varying from a few days to a couple of months, leading to consequent variation in brightness due to complex physical phenomena related to their evolutionary stage. The greater the period of oscillation, the greater their intrinsic brightness.

Henrietta Leavitt was one of Harvard's famous "computer women," astronomers hired by Edward C. Pickering to catalog the spectra and the brightness of the stars in the photographic plates of the observatory. It is estimated that in four years of work they cataloged 225,000 stars, publishing a first catalog in 1890, which was then extended to 360,000 stars in 1949. Leavitt's discovery was so important that in 1924 the Royal Swedish Academy had started proposing her for the Nobel Prize. Unfortunately, she had died three years before.

The importance of Cepheids is that they are *standard candles*, reference stars for which the *absolute magnitude* or *intrinsic brightness* can be determined, which in the case of Cepheids is linked to the period of oscillation. From the *intrinsic luminosity* and from the *apparent luminosity*, that is the observed luminosity, one can trace the distance.

To better understand, suppose we have a light bulb whose power we know, usually measured in watts, also called *brightness*, which is the amount of energy emitted by the light bulb in a second. If we put the bulb at increasing distances, we observe that it turns out to be increasingly dimmer. The intensity of the light we observe changes as the distance changes. Such an amount that we could call *apparent brightness*, or technically *flux*, depends on the intrinsic or real brightness of the bulb and on the distance and decreases with the square of the latter. In other words, if the flux of the lamp at 1 m is 1, at 10 m it will be 1/100 and at 100 m it will be 1/10,000. Knowing the intrinsic brightness of the bulb and measuring its apparent brightness, the flux, we can determine how far the bulb is from us using this known relationship. So, to determine the distance we must measure the flux and know the intrinsic brightness of the object. For a bulb that we buy, the brightness is already given by the manufacturer, but for an astronomical object it is not so. However, there are objects for which we know or can calculate the intrinsic brightness, the standard candles. To give another example, suppose that fishermen at night want to know how far they are from the coast and that they can see the light of a lighthouse. The further away they are from the coast, the dimmer the light will be. To have an exact measurement of the distance from the coast, the fishermen must have asked about the actual brightness of the lighthouse beforehand and measured the apparent one. To determine the distances in cosmology we need cosmic "lighthouses."

1.6 Hubble and the Expanding Universe

Although Lemaître had theoretically determined the law of expansion of the universe, due to the implications we previously mentioned, this had not been accepted by scientific circles. He himself had contributed to this, since as already mentioned, Lemaître himself in the English translation of his 1927 article had deleted almost all references to the expansion law.

Hubble's most important result was published in the article "A Relation between Distance and Radial Velocity among Extra-Galactic Nebulae" in 1929. He used the Doppler effect to determine velocities and Cepheids to determine distance, along with the distances and velocities of 24 galaxies obtained from Vesto Slipher,[10] not mentioned by Hubble, along with other partial information on 22 other systems. Hubble showed that there was a relationship between the recession velocities of galaxies and their distance: $v = Hr$, the famous *Hubble law*. The quantity H is known as

---

[10] Slipher observing spiral nebulae noticed, before Hubble, a shift of the spectral lines towards the red of the spectrum which implied a recession of the spiral nebulae. In practice he had found evidence of the expansion of the universe, but he did not realize it because he did not know that the nebulae were galaxies like ours

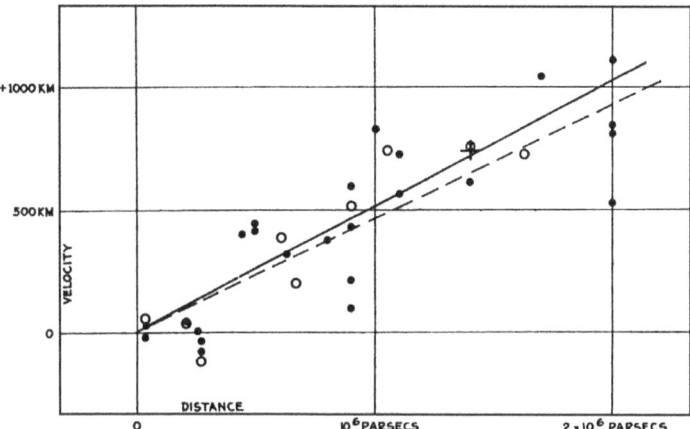

Fig. 1.2. Hubble law original data. Speed increases proportionally to distance between galaxies. Credit: Robert P. Kirshner (https://www.pnas.org/content/101/1/8) (Copyright 1929, The Huntington Library, Art Collections and Botanical Gardens).

the *Hubble constant* and represents the expansion rate of the universe. At that time, it had been erroneously estimated at 500 km/(s Mpc) (see Box 1, Distance Measurement Unit in Astrophysics). The original data of the Hubble law are represented in Fig. 1.2.

The error, as shown by Walter Baade in 1950, was mainly due to the fact that there are two different populations of Cepheids (Cepheids and RR Lyrae) with different luminosity period relationships and that the Andromeda Cepheids are four times brighter. For this reason, Hubble's estimate of the Andromeda distance (about 1 million light-years) was about a factor 2 lower than the real distance, known today to be 2.5 million light-years. The inverse of the Hubble constant gives an estimate of the age of the universe, which, with the value obtained by Hubble, provided only 2 billion years — much lower, for example, than the age of the Earth.

Using one of the values currently considered correct, 74 km/(s Mpc), we obtain an age of the universe of about 14 billion years, in accordance with other data.

From the data used by Hubble the linear relationship between speed and distance is not immediately visible, and doubts have been put forward as to how Hubble could have derived a linear relationship from such poor data. It is also unlikely that Hubble had realized the possibility of such a relationship from the study of previous works, such as that of Lemaître in 1927, as it is not known if Hubble knew about this study, written in French. In the 1931 English translation, the sections in which Lemaître established the speed–distance relation and the previous paragraphs in which he discussed the astronomical data and the redshift of the nebulae, used to obtain his relationship, were missing. In the article only one sentence was left: "starting from a discussion of the available data, we adopt $R'/R = 0.68 \times 10^{-27}$ cm^{-1}."

Lemaître never dedicated himself to promoting his results. Only on one occasion, in 1950, he returned to the subject by showing that he had no intention of giving up on his 1927 result. Posterity gave Lemaître the recognition he deserved. Given the role of Lemaître in determining the speed–distance law, an International Astronomical Union (IAU) resolution in 2018 changed the name of the Hubble law to the *Hubble–Lemaître law*. Here it should be added that for fairness, the law should, at least, also contain the name of Vesto Slipher, who was actually the first to observe the recession of the galaxies.

All in all, Hubble's law is the first fundamental test underlying the Big Bang theory.

There are two other pillars on which this theory is based and which we will discuss extensively in the next chapter. We only mention that the second is the *cosmic microwave background radiation* (CMB), discovered in 1965, which is a sort of residual microwave sea that pervades the universe and reaches Earth from all over the sky. A real imprint of the Big Bang. It was predicted in 1948 by Alpher, Herman, and Gamow, and its temperature, now known to be 2.725 K, is very close to the value obtained by Alpher and Herman in 1950.

The third is the predictions of the abundance of light elements, generated in the primordial nucleosynthesis, obtained with the calculations of Ralph Alpher and published in the famous article "$\alpha\beta\gamma$," which we will talk about.

The definitive acceptance of the Big Bang theory came with the accidental discovery of the cosmic background radiation. Besides Lemaître, who rediscovered Friedman's solutions and linked the expansion of the universe with observations of the redshift and the theory of the primordial atom (which presumed an initial state of the universe), Gamow, Alpher, and Herman are to be considered the fathers of the Big Bang model. Perhaps we should add Friedman who, according to Gamow, his student in Petrograd, was the first to speak of an initial state with very high temperatures that would later "explode": the Big Bang.

Since 1948, when Gamow, Alpher, and Herman made their calculations public, nearly 20 years had passed in which nobody cared about their results. Only nearly 20 years later, in 1964, Hoyle, Peebles, Zeldovich, and Tayler resumed the nucleosynthesis calculations already made by Gamow, Alpher, and Herman.

This time lapse of "silence" was perhaps due to the fact that neither Gamow nor his students fully realized the importance of their results. We should add to this the fact that Alpher and Herman's discussions with radiation experts led to nothing, as it seemed impossible to reveal radiation below 10 K with the technologies of the time. Yet it has not only been revealed but has become one of the fundamental foundations of the Big Bang theory, and a bottomless well from which to obtain information on the primordial universe.

1.7 How Big and Old Is the Universe?

Although the theory has the name of Big Bang, it was not an explosion like the ones we know, where the explosion starts from one point. The Big Bang happened everywhere. A simple example that clarifies the expansion of the universe and Hubble's law is that of a balloon that inflates and on which dots are drawn (Fig. 1.3). In the analogy the dots represent the

The Invisible Universe

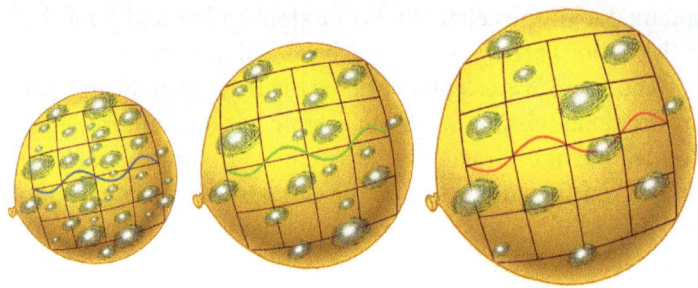

Fig. 1.3. During the expansion of the universe, represented by the balloons, the galaxies move away. With expansion, the wavelength of radiation is stretched due to space stretching.

galaxies, and the balloon the space. When the balloon inflates, the rubber is stretched and the dots on it move away from each other. If we suppose ourselves to be one of the dots, we see all the other dots moving away from us and we think we are at the center of the universe. In fact, we are not standing still, but we are moving away from the other dots. From any point the observed situation is identical: the other points are moving away with a speed proportional to the distance. In other words, there is no privileged point. The expansion is identical for every point in the universe. As shown in Fig. 1.3, with the expansion of the universe, the wavelength of light increases; it is "stretched."

Another analogy often used is that of bread dough, which represents space, containing raisins, which represents galaxies. During the leavening phase, the dough increases in volume and the raisins, although not moving, move away from each other because the dough expands and drags them with it (Fig. 1.4).

An important point to note is that the expansion of the universe does not concern objects such as planets, stars, or galaxies, since on this scale the gravitational force dominates the expansion.

Hubble's law also allows us to know the age of the universe and its size.

To estimate the age of the universe, suppose that the rate of expansion has always been the same throughout the life of the

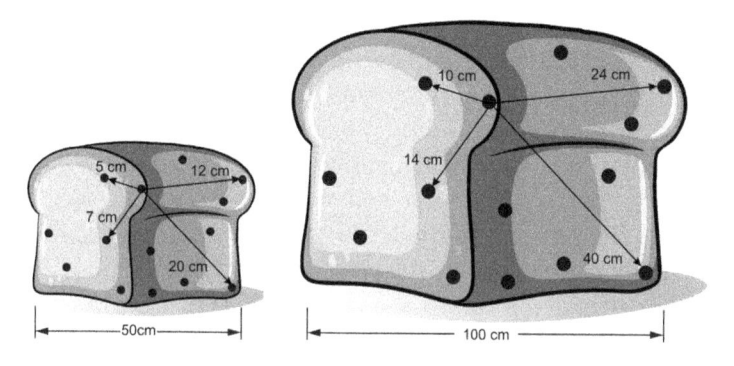

Fig. 1.4. Example showing how universe expands.

universe. If the expansion of the universe occurs quickly, the universe is expected to be young, and vice versa. The Hubble constant tells us how fast the universe expands, and this allows us to determine its age. If we reverse the direction of time, the galaxies that are moving away today would begin to approach, ideally, to join in one point. So, if two galaxies are separated by a distance D today, and v is their separation speed, the time required for them to reach separation D is given by the relationship between distance and speed. Recalling the Hubble law, the age of the universe is the inverse of the Hubble constant. If we use the H value ~74 km/(s Mpc) obtained in 2019 by the Hubble Space Telescope,[11] we obtain an age of 13.6 billion years versus the more precise experimental value of 13.8.

As for the size of the universe, we know that it is geometrically flat and therefore infinite in accordance with the inflationary paradigm.[12] In practice, things are a little more complicated. If we now want to know what is the size of the universe that we can access, we must remember that in order to observe an object we must receive its light. Since the universe has a measured age of 13.8 billion years, the objects we can observe cannot be more distant than the space

---

[11] In 2018 the Planck mission achieved a lower value, ~67 km/s/Mpc.
[12] The rapid expansion that took place at the origin of the times that brought the universe from sub-atomic dimensions to those of a soccer ball. See the Thermal History section of Chapter 2, and Appendix A.

that light can have traveled in that time. Consequently, there is a *cosmological horizon* beyond which we cannot see. The *observable universe* constitutes only a part of our universe.

So even if the universe is infinite, we still have access to only a part of it, and the dimensions of our universe are those of the universe observable today, which tomorrow will be larger. The situation is similar to that of a sailor on a ship in the middle of the ocean. The view is limited from the horizon of the ocean. If the sailor had a way to be able to move upwards from his ship, the horizon would widen, as does the horizon of our universe over time.

An appropriate estimate of the size of the universe requires the use of general relativity. This estimate says that the size of the visible universe is about 46 billion light-years, or about 20,000 times greater than the distance between us and the nearest galaxy, Andromeda.

Ultimately, the question "How did the universe originate?" which has puzzled philosophers and scholars since the beginning of time, today has its answer in the Big Bang theory: the universe was created in a singularity that created not only matter but also space, time, and the laws that govern our cosmos.

Obviously, cosmologists do not sit idly and alternative theories to the Big Bang have been proposed. As we will see in the last chapter, some cosmologists are convinced that the Big Bang does not correspond to the beginning, but to a "transition door" between two stages. The Great Explosion could have happened after a previous *Big Crunch*, followed by a *Big Bounce*, followed by a new expansion. This scenario could repeat itself periodically giving rise to a cyclic universe. Another possibility, avoiding the singularity, is that based on *string theory*,[13] whose physics does not allow the dimensions of the universe to contract to dimensions smaller than those of Planck. For this reason, when the collapse reaches this dimension, the universe undergoes a Big Bounce. Even in *loop quantum theory*, which

---

[13] As we will see in Chapter 7, string theory assumes that the particles that make up the universe are generated by the oscillation of tiny strings.

seeks to create a single theory from quantum mechanics and general relativity, the Big Bang does not exist.

There is another non-singular model based on *M theory* that combines the 5 different string theories. According to the *ekpyrotic model* (the term coming from the Stoic philosophy and that can be translated as "transformation into fire" or "out of the fire") of string theory, our universe could be found on a three-dimensional *brane*, a sort of membrane that floats in a multidimensional space. The Big Bang would have originated from the collision of our brane with another, once or cyclically (see section 10.5). Finally, some physicists have tried to eliminate the singularity, namely the Big Bang.

For example, Stephen Hawking proposed the concept of *imaginary time*,[14] which behaves like a fourth spatial dimension. Its use in cosmology can eliminate singularities, such as the Big Bang. By removing singularities, the universe becomes borderless. We will talk more about these topics in the final chapter.

In conclusion, although the Big Bang is not the only theory that describes how the universe originated, it is certainly the one with the most data supporting it, the most accepted and popular.

---

[14] The concept of imaginary time is linked to the so-called Hartle–Hawking state, the wave function of the universe, a proposal on the state of the universe before the Big Bang.

Chapter 2

THE PRIMORDIAL UNIVERSE

There is no darkness but ignorance.

— William Shakespeare

2.1 Rewinding the Movie

Today we are fairly certain that our universe originated 13.8 billion years ago and was a very different place from what it is today. Today it is expanding at a speed of about 74 km/s/Mpc (for the definition of megaparsec, Mpc, see Box 1: Distance Units in Astrophysics). If we think of the universe as a movie, we could rewind it and we would expect to see the galaxies reverse their motion and begin to move towards a point. The energy of the photons is proportional to their wavelength (see Box 2: Light and the Electromagnetic Spectrum), and smaller wavelengths correspond to higher energies. Since the wavelength is *stretched* by the expansion and compressed in the collapse phase (see Box 3: Doppler Effect), the energy of the photons increases and the temperature does the same. This means that in the past, in addition to being smaller, the universe was also hotter. Going far back in time, temperatures were so high that matter as we know it today did not exist. Atoms have binding energies of a few tens of eV (see Box 1: Temperature and Energy Units), while nuclei of the order of millions of eV.[1] At higher temperatures, atoms and nuclei could not exist as related objects. Even if we could make

---

[1] Translated into temperature, the indicated energy corresponds to ten billion degrees K (10^{10} K).

a journey to the origin of the cosmos, the laws of known physics would allow us to understand the universe only up to scales greater than Planck's, about 10^{-35} m. On smaller scales, a theory that still does not exist is needed, namely *quantum gravity* which is a combination of gravity theory and quantum mechanics. In practice, even having formulated this theory, we could understand the universe only from a theoretical point of view, but we could not observe it with electromagnetic radiation, since from the Big Bang to 380,000 years, the matter in the universe was so dense that it was completely opaque. There is a sort of impenetrable wall between us and the Big Bang. Starting from Planck's time, known physics allows us to describe the most important phases of the evolution of the universe. Now we will talk about it a little.

2.2 Thermal History

About 14 billion years ago the universe had infinitesimal dimensions and contained all the matter and energy that make it up today. According to the modern idea, it was created from nothing. This idea clashes with religious thought and common sense, according to which "nothing is created from nothing," as Parmenides of Elea claimed already in the sixth century BC. In reality, when we talk about nothing, we refer to the *quantum vacuum* (see Box 5 and section 6.2).

According to modern physics, the universe could be created from nothing, as its energy, given by the sum of the positive energy of matter and the negative energy of the gravitational fields, is zero! This is in accordance with the *inflation theory*,[2] and with the fact that the universe has a flat geometry. The universe was generated from nothing, where, I repeat, for nothing we must understand the *quantum vacuum*. This entity does not correspond to the common notion of emptiness as the absence of everything, but to the minimum of energy of quantum fields.

---

[2] Exponential expansion phase of the universe in its initial phase (see Appendix A).

The Primordial Universe

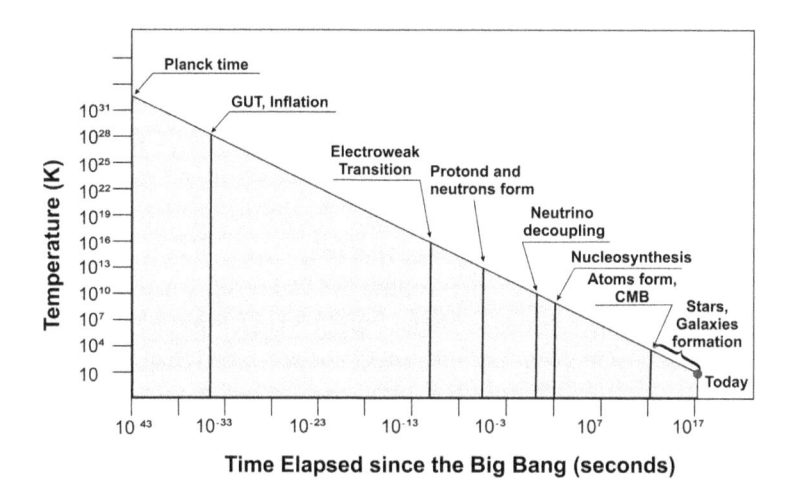

Fig. 2.1. Temperature in the primordial universe.

The first moments after the Planck epoch (10^{-43} s) are characterized by pressures, temperatures (10^{33} K [Fig. 2.1]) and densities (10^{96} g/cm^3) so high that the space was distorted, folded and constituted by a foam of mini black holes, *wormholes* (sort of gravitational tunnels) with dimensions of the order of 10^{-35} m, temperatures of 10^{32} K, and evaporation times of the *Planck time* order. The cosmological horizon of the universe was about 10^{-35} m.

These are such extreme conditions that cannot be re-created even in the most powerful particle accelerator in the world, LHC. To do this, in the estimates of Casher and Nussinov,[3] an enormous accelerator would be needed from about 30 parsecs to 30 kiloparsecs, i.e., equal to the size of our galaxy. In today's universe, gravity is much less intense than other forces. On cosmological scales, gravity triumphs and we have a theory for it, general relativity. On the atomic scale, the other forces dominate and we have another theory that describes the world, quantum mechanics. However, in Planck's time, the gravity that in our world is enormously less intense than the other forces, has intensity comparable to

---
[3] https://cds.cern.ch/record/290203/files/9510364.pdf

the others. We need a theory that brings together quantum mechanics and general relativity, *quantum gravity*, which, as we have already said, has not yet been formulated. The four forces of nature were unified and mediated by the so-called *X and Y bosons* (see Box 4). We cannot penetrate the secrets of the period from the Big Bang to the *Planck era* simply because we do not know the physics of that era.

So our physical theories allow us to start our story starting from the *Planck era*. Due to the finiteness of the speed of light, an observer in *Planck's time* could not see more than 10^{-35} m away, a factor 10^{20} smaller than the size of a proton. The universe is dominated by the quantum vacuum (see Box 5),

---

BOX 4

SUMMARY: ELEMENTARY PARTICLES AND INTERACTIONS

Here we recall some notions relating to particles and elementary forces described in Chapter 6.

Spin: Spin was introduced in 1925 by Goudsmit and Uhlenbeck to explain previous experiments. They suggested that electron had an *intrinsic angular momentum*, that is, it was rotating around its own axis like a top. Even if not correct, this idea is often used in popular books. Spin, as shown by Dirac, comes out automatically in the relativistic version of quantum mechanics. It is a quantum number that together with other quantum numbers describe the quantum state of a particle. The value of the spin of a particle describes to which family the particle belongs, as we will see in the following.

Particles are divided into two groups, *fermions* and *bosons*.

Fermions: particles that constitute matter. They are characterized by a half-integer spin: ½ ℏ, 3/2 ℏ ...

where ℏ is a number called *reduced Planck constant*.

Examples: *quarks* (which constitute protons and neutrons), *electrons*, *muons*, *tau*, *neutrinos*.

Bosons: particles that mediate forces. They have an integer spin: 0 ℏ, ℏ, 2 ℏ ...
Examples: *photons, W⁺, W⁻, Z⁰ bosons, gluons*.

1. *Hadrons*: They are fermions. They are particles subject to a strong nuclear force and can be divided into

   - *Baryons*: These are fermions subject to a strong nuclear force. The baryons are not elementary particles, but made up of elementary particles called quarks. Examples of baryons are *protons* and *neutrons*. Outside the nucleus, neutrons are unstable and have an average life of about 15 minutes. Protons are almost immortal particles, given that their average life of 10^{34} years is much greater than the age of the universe.

BOX 4 (*Continued*)

- *Mesons*: They are bosons, subject to a strong nuclear force. They consist of a quark and an antiquark.

2. *Leptons*: they are fermions. They do not "feel" strong interaction, they are elementary particles, and are fermions with spin 1/2. Examples of this family are charged particles such as *electrons, muons, tauons*, their respective antiparticles, and *neutrinos*. The latter are presented in 3 flavors: *electron neutrino, muon* and *tau* and can oscillate, that is, transform into each other, in the phenomenon of *neutrino oscillations*.

In nature there are *four interactions* or forces: *gravitational, electromagnetic, strong nuclear* and *weak nuclear*. The standard model of particle physics only describes the last three.

- *Electromagnetic force*: it can be attractive and repulsive and its intensity decays with the square of the distance. The electromagnetic force is mediated by the photon.
- *Strong nuclear force*: its intensity grows with decreasing distance, and it is responsible for the existence of the atomic nucleus. It has a radius of action of the order of the size of the proton. It is mediated by 8 bosons, dubbed *gluons*.
- *Weak nuclear force*: responsible for radioactive decay. It has three mediators, the bosons W^+, W^-, Z^0. The weak nuclear and electromagnetic forces are manifestations of a single force, the *electroweak forces*.
- *Gravitational force*: it is always attractive. It has a theoretical mediator, never observed, the *graviton*.
- The *Higgs boson* supplies mass to elementary particles such as electrons, quarks and bosons W^+, W^-, Z^0

BOX 5

QUANTUM FLUCTUATIONS

In the book, we will often talk about *quantum fluctuations*. What do we mean by this term?

To understand this point, we must introduce *Heisenberg's uncertainty principle* of quantum mechanics, which requires that some physical quantities, called *conjugate variables*, such as *energy* and *time* or *position* and *moment*, (i.e., the product of mass by speed) are subject to a certain degree of uncertainty. In other words, in a physical system it is not possible to determine with infinite precision all the values of the physical observables at the same time, unlike in classical physics. According to Heisenberg's uncertainty principle, the more precisely the position is measured, the less we know about the moment, and this also applies to the couple energy, E, and time, t. The important thing to remember is that this principle does not depend on the precision of the experiments. Even in an ideal and perfect experiment, uncertainty would continue to exist. So not only is it not possible to know the exact energy of a system in a certain instant, but if we wanted to determine exactly

(*Continued*)

BOX 5 (*Continued*)

the energy of the system we would need infinite time. As a consequence, a system does not have a defined energy at all times, but it changes, fluctuates permanently, that is, its value increases and decreases compared to what we would expect, and all this at a speed that cannot be measured directly. In other words, the conservation of energy can be violated, but only for very short periods of time: the lower the energy present in the fluctuation, the longer it can persist. Quantum uncertainty allows small quantities of energy to appear out of nowhere, provided that they disappear in a very short time. This energy can take the form of pairs of particles and antiparticles, called *virtual particles*, of very short life — for example, an electron–positron pair. Therefore, Heisenberg's uncertainty principle causes the more uniform environment to necessarily have quantum irregularities.

WHAT IS A FIELD?

As we will see in more detail in Chapter 6 (section 6.2), a field is a function that assigns a value to every point in space. For example, the values of temperatures associated with the spatial points of a room constitute a field, called a *scalar field*. The set of vectors that describe the movement of air in a room constitutes another field, a *vector field*. However, these fields are not fundamental fields; they are air properties. Examples of fundamental fields are the *gravitational* and *electromagnetic fields*. The quoted fields are classic fields. If we apply quantum theory to fields, the result is a quantum field, subject to the rules of quantum mechanics. Each field has its own energy and associated particles. For example, *photons* are associated with the *electromagnetic field*. More precisely, the *particles are nothing but fluctuations in the field, and the fields are the real building blocks that make up the universe*. Furthermore, the quantities characterizing a quantum field are subject to *fluctuations*, due to the *uncertainty principle*. The energy in a point constantly fluctuates, and in the field *virtual particles* appear and disappear.

THE QUANTUM VACUUM

In physics, every system tends to reach the state of minimum energy. The *vacuum is the minimum energy state of a quantum system*. As a consequence of what has been written and of the uncertainty principle, the quantum vacuum is not completely empty as we imagine it. The quantum vacuum is a sort of sea full of virtual particles that are born and annihilated in a very short time. So the emptiest region we can imagine seethes with particles, and the energy itself of the region is constantly changing. A more complete answer referring also to the concept of field is given in section 6.2.

from which particle–antiparticle pairs are created, and return to *vacuum* after infinitesimal times.

The *Grand Unification Age* follows, which unfolds between 10^{-43} s and 10^{-36} s. At the beginning of this epoch the force of gravity separated from the other forces described by the *Grand Unification Theory* (GUT) mediated by the hypothetical bosons X and Y. The temperature of the universe decreased with expansion, and in this phase of the universe the decrease was very rapid. Between the *Planck era* and the end of the Grand Unification era, the temperature dropped to temperatures below 10^{30} K. At the end of this era, when the temperature reached 10^{28} K, the universe was subject to the *Great Unification phase transition*, which led to the decoupling of the strong interaction from the electroweak force, constituting two separate forces. Some types of phase transitions are present in everyday life. The cooled water vapor turns into water at 100°C, and at temperature smaller than 0°C water turns into ice. In these transformations, heat, called *latent heat*, is released. In addition, we move from higher-symmetry to lower-symmetry systems. For example, water is the same from whichever direction we look at it, while ice has a crystalline structure and is not as symmetrical as water. In Guth's *inflation theory*, the grand unification phase transition gave rise to cosmic inflation, a period of exponential expansion, which will be discussed in a moment and in more detail in Appendix A. Due to the phase transition, a huge amount of energy was released in the universe, similar to what happens in the water–ice phase transition, which releases latent heat, as mentioned. The universe, of sub-nuclear dimensions, underwent a metamorphosis, thanks to inflation, which developed in the period between 10^{-36} and 10^{-32} s. From dimensions of the order of 10^{-28} m it grew by a factor of 10^{25}–10^{30} (see Appendix A), or perhaps higher. In an infinitely small time the universe reached the size of an soccer ball. This huge expansion produced a collapse in temperature. The universe was super-cooled from 10^{27} to 10^{22} K and then heated again in the *re-heating phase*. The release of the energy of the vacuum transformed the virtual particles, which appeared and disappeared in Planck's time, into real particles. In subsequent versions of inflation, for example Linde's inflation, the expansion is produced by the decay of the potential energy of a field called *inflaton* (see Box 5 and Appendix A). The universe was filled with radiation and elementary particles:

quarks, leptons and their respective antiparticles. These events occurred between the GUT era and an "extremely" long era, called the *electroweak era* that developed in the period between 10^{-36} and 10^{-12} s.

In this era, the universe was full of particles and antiparticles that began to collide and then annihilate, leaving a surplus of matter particles that would then form our world. In reality, one would expect that all particles and antiparticles would annihilate, but this did not happen due to a very small asymmetry between the number of particles and that of the antiparticles (we will talk about this in a moment). Hypothetical components of cold dark matter (e.g., *axions*) may have formed in this era. Some theorists have assumed that the agent that produced inflation was the *Higgs field*, a scalar field (a field with zero spin) that, interacting with elementary particles such as quarks, W, Z particles, and leptons, provided it with mass (see section 6.9 and Appendix C).

Inflation is a particularly important phase in the evolution of the universe. Although this phase is not yet well known, as demonstrated by the fact that there is not a single inflationary theory but several, it allows to explain many otherwise inexplicable things. Inflation solves a number of problems in the Big Bang cosmology. It explains the homogeneity of space, provides the universe flatness in accordance with the observations of the *cosmic background radiation*[4] (CMB) (of which we will speak more in the following), and provides the generation of small inhomogeneities from which the galaxies were formed.[5] Finally, it produces a *primordial gravitational wave background*,[6] that is, waves produced by the motion of

---

[4] The cosmic background radiation, as already mentioned, is a sort of residual microwave sea that pervades the universe and reaches the Earth from all over the sky.

[5] Technically one speaks of an *initial spectrum of perturbations* generated by quantum fluctuations (see Box 5).

[6] Similarly to a charged body that emits electromagnetic waves, a massive one emits gravitational waves. Predicted by Einstein in 1916, they were searched for decades and discovered only in 2015 thanks to the LIGO/Virgo collaboration. The first event revealed was that of two black holes, 1 billion

massive bodies (even if not charged). The initial idea that gravitational waves could be produced by quantum fluctuations (see Box 5) in an expanding universe, was proposed by the Ukrainian physicist Grischuk in 1975 and later by Rubakov, Sazhin, and Veryaskin in 1982, Fabbri and Pollock in 1983, and Abbot and Wise in 1984.

Finally, in the new versions of inflation, for example Linde's, inflation does not only create our universe, but creates infinite ones, the so-called *multiverse* (see Chapter 10 and Appendix A).

Although there is a large group of physicists studying inflation, eminent physicists criticize it — among them, even Paul Steinhardt, who was one of its founders. As we will see in the last chapter, he proposed an alternative universe model, called the *ekpyrotic universe*, and an evolution of this called *phoenix universe*.

Returning to our story, at that time, the energies were so high that two colliding photons formed particles of matter that annihilated encountering antimatter particles. In short, there was a continuous flow of creation and destruction. At the end of the electroweak era, the W and Z bosons (see Box 4) were no longer created and decayed quickly. The electroweak nuclear force became a short-range force separating into electromagnetic force and weak force. The universe underwent a second phase transition. This transition is indicated with the term *electroweak symmetry breaking*.

In the first billionth of a second (10^{-12} s), the universe was made up of a *plasma of quarks*, leptons, and gluons, and its dimensions were approximately equal to the distance from Earth to Sun. This quark soup, called *quark–gluons plasma*, was first observed at the Brookhaven National Laboratory in 2002. Over time, the energy of the quarks decreased and around 10^{-6} s, when the temperature went below 1000 billion degrees, the universe was no longer able to keep apart

---

and 300 million light-years away, which in a dance that was at first slow and then faster and faster, until it reached speeds close to that of light, merged, generating a single black hole (cf. section 5.5)

the quarks that formed *hadrons*, which are a large family containing the *baryons* (e.g., protons, neutrons) containing an odd number of quarks and the *mesons* consisting of two quarks (see section 6.9).

Protons are almost eternal particles, while neutrons decay in 15 minutes. Meanwhile, the temperature continued to drop inexorably. This era is called the *quark era*. It extends between one thousandth of a billionth and one millionth of a second (10^{-12}–10^{-6} s).

Particles and antiparticles were converted into photons, but the latter were no longer energetic enough to create antiparticle pairs. The universe was dominated by photons, electrons, neutrinos, and a minority (one part in a hundred million) of protons and neutrons.

When the temperature dropped below 10^{12} K this era, dubbed the *hadronic era* (10^{-6}–1 s) (when the quarks were confined to form hadrons), ended. The *leptonic era* (1 s–3 min), dominated by leptons and anti-leptons, followed. Electrons and positrons collided and annihilated generating photons, which in turn collided forming other electron–positron pairs. The decrease in temperature and density produced a decrease in the reaction rates with the consequent decoupling of neutrinos, when the universe had 1 second. After photons, they were the largest population in the universe. They formed a background, the *cosmic background of neutrinos*, with a current temperature of 1.95 K, slightly lower than that of photons (2.725 K). Although there is indirect evidence of the existence of the *neutrino background*, it has never been detected, given the poor interaction of neutrinos with matter. Electrons and positrons annihilated, leaving behind only particles of matter.

The Big Bang created an equal amount of matter and antimatter. The particles of matter and antimatter are created in pairs and have the same mass but opposite charge. If they meet, they annihilate, leaving only energy behind. But today's universe seems to be predominantly made up of matter. Where did all the antimatter go?

The origin of matter is one of the greatest mysteries of physics. To solve this problem, it is assumed that there was a slight asymmetry in the number of particles of matter and antimatter: an additional baryon for every billion anti-barions. Andrei Sakharov explained this asymmetry by proposing three conditions that every interaction that produces baryons must satisfy. In physics there are different quantities that are conserved like energy, and among these the number of baryons. The first of the assumptions is precisely that this number is not preserved. The second is that the interactions do not take place in thermal equilibrium, and finally that the production of matter is privileged over that of antimatter, that is, it violates the symmetry of charge and parity, called *CP symmetry*. *CP symmetry* is composed by two laws of symmetry: *C symmetry* (standing for *charge symmetry*) and *P symmetry* (standing for *parity symmetry*). C symmetry is a law of symmetry related to the exchange of a particle with its antiparticle, for example, exchange of a positron with an electron. P symmetry is a law of symmetry related to the inversion of the spatial coordinates: (x, y, z) to $(-x, -y, -z)$.

Going back to our story, the annihilation of most electrons has big effects on that of neutrons. In fact, in the *hadronic era*, neutrons did not disappear, because they were formed by the fusion of protons and electrons with the production of neutrinos. Following the annihilation of electrons, protons could no longer generate neutrons, whose number decreased. When the universe was a little older than 1 second, for every 10 protons there were only 2 neutrons. Neutrons are a little heavier than protons and decay in about 15 minutes. When the temperature was around 10^{10} K, neutrons were no longer created and the neutron–proton ratio was frozen to about 1 neutron for every 10 protons. Due to the decay of neutrons, their number, at 300 seconds, when the temperature had dropped to about 10^9 K, had decreased to 2 against 14 protons. The number of neutrons and protons was sufficient to form 1 helium-4 nucleus, leaving 12 protons for each helium-4 nucleus. Since a helium nucleus is four times heavier than a hydrogen nucleus, the helium-4 fraction was given by 4 divided by 12 + 4 (16), and therefore in the universe there was about 25% of helium and the 75% of hydrogen.

In more detail, when the temperature dropped below a few billion degrees, protons and neutrons generated stable nuclei such as deuterium, which does not have a long life. Things changed when the universe was about 3 minutes old. Deuterium no longer decayed and helium-4, helium-3, and lithium-7 were formed. We are in the *era of nucleosynthesis*, between 3 seconds and 20 minutes. As seen, the abundance of hydrogen (75%) and helium (25%) formed in the

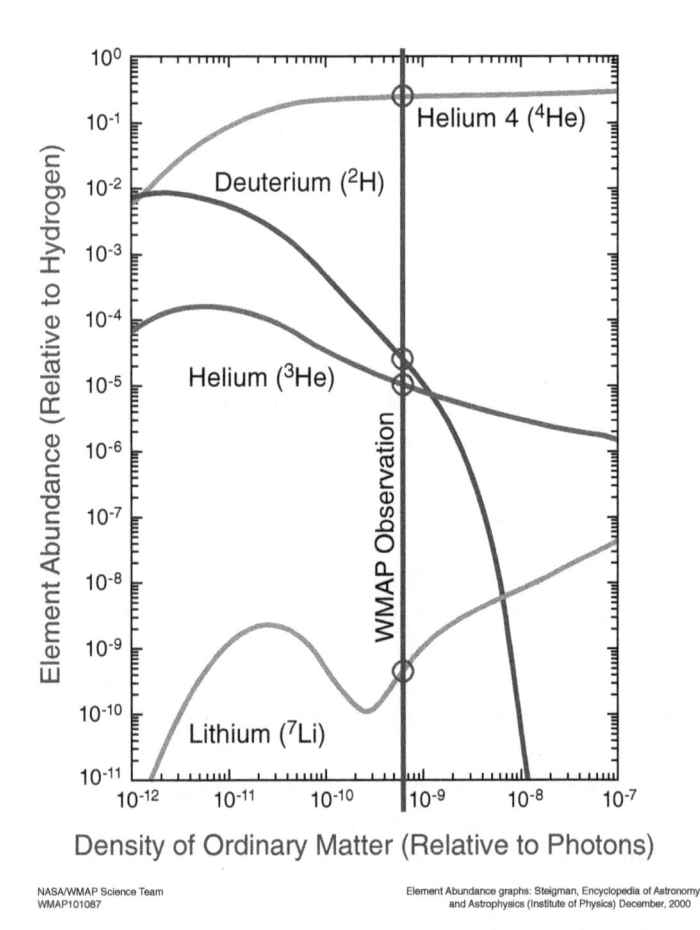

Fig. 2.2. Relative abundance of light elements in primordial nucleosynthesis. Credit: NASA/WMAP Science Team.

nucleosynthesis depended on the density of protons, neutrons, and radiation (Fig. 2.2).

These numbers have been confirmed by the measure of the abundance of the elements formed at that time, having remained unchanged until today, despite the transformation of hydrogen into helium in the stars. After about 20 minutes the temperature dropped to the point where nuclear fusion stops.

The first nucleosynthesis calculations were made in the 1948 by Ralph Alpher and George Gamow,[7] and published in the famous article "αβγ," from the initials of Alpher, Bethe, and Gamow. The calculations had been published in Alpher's thesis, but Gamow believed they deserved wider circulation and published an article in the journal *Physical Review*. Given his sense of humor, he added the name of Hans Bethe to the paper for the reason he explained in this way:

> It seemed unfair to the Greek alphabet to have the article signed by Alpher and Gamow only, and so the name of Dr. Hans A. Bethe (in absentia) was inserted in preparing the manuscript for print. Dr. Bethe, who received a copy of the manuscript, did not object, and, as a matter of fact, was quite helpful in subsequent discussions.

Bethe accepted and contributed to the article that appeared on April 1, 1948.

Primordial nucleosynthesis produces only light elements, up to beryllium. Since there is no stable nucleus with

---

[7] Gamow was a Ukrainian physicist, born in Odessa. An eclectic character with a great sense of humor, he dealt not only with cosmology and nuclear physics, but also with molecular biology: he made a contribution to deciphering DNA. When he was 18, he went to study at the University of Novorossiysk and then to Leningrad (present-day St. Petersburg) with Aleksandr Friedman. Between 1928 and 1931 he studied in Copenhagen at the Bohr Institute of Theoretical Physics, and at the Cavendish Laboratory. After a period in Göttingen, where he made a great contribution to nuclear physics, he was called back to Leningrad as a professor, but he definitively left Russia for America because of the Stalinist regime. He was also a great popularizer.

8 nucleons, the nucleosynthesis stops. Another fundamental prediction of Gamow, together with Alpher and Herman, was that the universe had to be immersed in a microwave radiation background, the famous cosmic microwave background (CMB), with a temperature of 5 K. Although this story is not well known, in 1950 Alpher and Herman managed to improve this estimate, obtaining the value 2.8 K, a value remarkably close to that currently known (2.725 K). This prediction was forgotten until the 1960s, when Penzias and Wilson accidentally discovered CMB.

At the turn of the nucleosynthesis era, between 3 minutes and 240,000 years, comes the *era of radiation*. The universe contained plasma, an incandescent and opaque soup of protons and electrons. After the annihilation of leptons and anti-leptons, the energy of the universe was dominated by photons that interacted with protons, and electrons. Because of these continuous reactions, light underwent continuous deviations and reflections and was therefore trapped in the plasma. The universe was opaque and dark. The situation in the universe was very similar to that when we find ourselves immersed in a cloud or in a very thick fog. Fog is a cloud that forms in contact with the ground and is made up of water and ice droplets. If we are in a car, the light of the headlights cannot move freely because photons in their motion will hit the droplets of water and will be reflected towards another droplet and so on, remaining trapped. If the number of droplets decreases, the photons will be freer to move, such as when the fog disappears, they are able to move freely. This is what happened in the following phase, between 240,000–380,000 years, called the *era of recombination/decoupling*. When the temperature of the universe decreases, what happens is similar to when the density of the droplets of fog decreases.

Zeldovich, Kurt, and Sunyaev in 1967 and Peebles in 1968 studied the physics of recombination. The first element that captured electrons was helium. The first electron was captured when the universe was 18,000 years old, and its second when it was around 130,000 years old. Since helium is made up of two protons, it became neutral. Then came the turn of hydrogen, which contains only one proton. In the period between 260,000 years to 380,000 years (temperature 3000 K), protons captured electrons forming neutral atoms. The phenomenon is known as *recombination* (of electrons and nuclei), associated with the *decoupling* between radiation and matter. As the free electrons decreased, the photons trapped by the interactions, with the electrons, became free to move and reached us and the "fog" that enveloped the universe disappeared and it became transparent while a cosmic background of visible light was released. This fossil background, due to the expansion of the universe, is now observable in microwaves and is precisely the CMB.

The photons that were released at the time of recombination were very energetic compared to those observed today, while their number was much greater than that of the baryons. Today the number density of photons is about 413 photons per cubic centimeter, while there are only 2.5×10^{-7} baryons per cubic meter. Photons dominate baryons in number, their

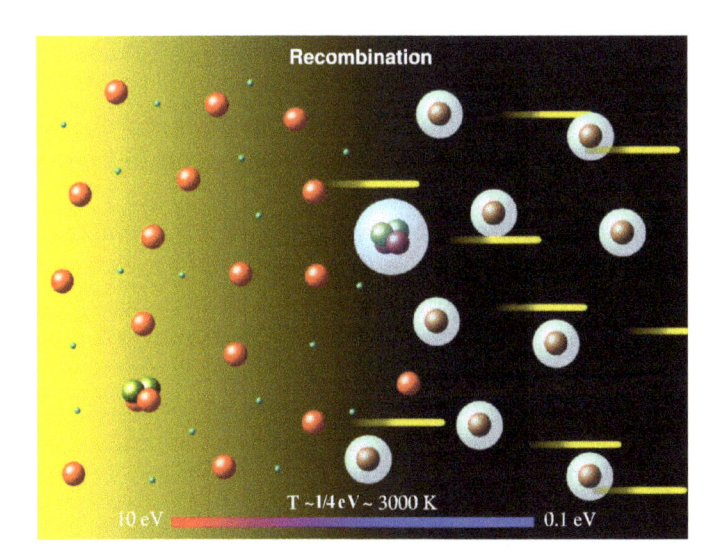

Fig. 2.3. Recombination. On the left, when the temperature was 10 eV the protons were not tied to the electrons, while at temperatures of 1/4 eV they formed the atoms. Credit: W. H. Kinney, "Cosmology, Inflation and the Physics of Nothing."

ratio is in the order of 6×10^{-10} (a few million photons for each proton), and their proportion has been preserved to the present. Due to the presence of this large number of photons, recombination related to hydrogen occurred later. In fact, an electron can combine with a proton if its energy is less than 13.6 eV, corresponding to a temperature of 158,000 K. This is the temperature at which the recombination related to hydrogen should have taken place. Instead, due to the large number of photons present, recombination occurred much later, at 3000 K. During recombination, not all electrons were coupled to protons. About one in a thousand remained free. Photons and free electrons continued to interact for about 9 million years and their temperatures remained connected until that time. This had significant implications on the formation of structures.

After the formation of atoms, practically nothing happened for hundreds of thousands of years, apart from the fact that the universe continued to expand and consequently cooled.

2.3 The Cosmic Background Radiation (CMB)

When the universe expanded, it cooled down, and the expansion "stretched" the wavelengths, increasing their length and decreasing their energy.

The temperature decreases in an inversely proportional fashion to the size of the universe, and the wavelength of photons increases with the size of the universe. Since energy is inversely proportional to wavelength, photons will see their energy halve as the size of the universe doubles.

The expansion from the time of recombination to now has enlarged the universe by about a factor of a thousand, similarly to the wavelengths. So the photons of the background radiation are today a thousand times less energetic than at recombination. Similarly, the temperature from 3000 K by a factor of 1000, reaching the value 2.725 K corresponding to a wavelength of 7.35 cm, and consequently it is nowadays observed in the spectral region of the microwaves. The radiation that has the same temperature in all directions of the sky is dubbed *cosmic background radiation*, or CMB, and is considered as the fossil residue of the radiation emitted at the time of recombination and the Big Bang itself. It is also possible to show that the temperature value depends on the distance. Considering galaxies closer to recombination, the temperature of the CMB in their surroundings is expected to be higher than what we measure. This temperature can be measured using cyanogen (CN), which acts like a microwave-sensitive thermometer. Indeed, it has been used to see that the temperature of the CMB at those distances is greater than that measured in our neighborhood.

This radiation was observed by chance in 1965 by two engineers, Arno Penzias and Robert Wilson, who were using a Bell telephone antenna to get the signal from the satellite Telstar.

They discovered microwave radiation that did not change with the time of day or orientation. Not having a preferential direction, it could not come from the nearby New York, or from the Sun. Initially they thought it was produced by some interference or by the "white dielectric material," that is, excrement, of a pair of pigeons, as Penzias called it, which were housed in the antenna.

They consulted an MIT astronomer, Bernie Burke, who knew about the studies of Robert Dicke and Jim Peebles from Princeton University. Those scientists were planning an experiment to measure background radiation, but were preceded by the two engineers. Penzias and Wilson published a short article in the *Astrophysical Journal* about what they had observed and in 1978 they won the Nobel Prize. In the previous pages of the same issue of the magazine, Dicke, Peebles, and collaborators explained the origin of the signal.

Neither Dicke nor Peebles were awarded the Nobel Prize at that time, but part of the 2019 Nobel Prize was awarded to Peebles for his contributions to cosmology. Returning to our theme, the discovery, together with the expansion of the universe, and the abundance of light elements in the cosmos, was another confirmation and completion of the Big Bang theory. The radiation was evidence that the universe had been in a hot phase, confirming the ideas of de Sitter, Lemaître, Gamow, and other scholars.

In 1969 it was discovered that this background radiation is not perfectly isotropic. There is one direction from which photons arrive with less energy, and are red-shifted, while in the opposite direction more energetic photons arrive, and are blue-shifted. This phenomenon, known as *dipolar anisotropy*, betrays the motion of our galaxy, with a speed of several hundred kilometers per second with respect to the CMB, towards the constellation of Virgo (Fig. 2.4), also moving towards the Hydra–Centaurus supercluster.

This motion is induced by large concentrations of masses in the surrounding environment, in a radius of 6 billion light-years, in which there are structures such as the *Shapley Attractor*, the *Great Attractor*, the recently discovered structures such as the *Super Cluster of Laniakea* (from the Hawaiian "immeasurable paradise") made up of 100,000 galaxies and identified in 2014 by the Brent Tully group of the University of Hawaii, and the *Dipolar Repulsor*, recently discovered. By subtracting the effects of motion, the homogeneity of the background radiation is regained, apart from the "imperfections" produced by the radiation of the galactic plane.

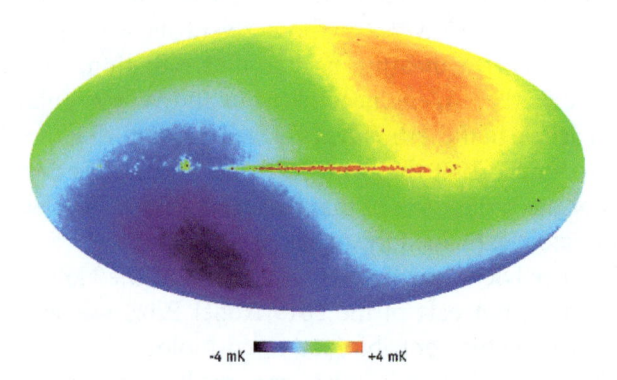

Fig. 2.4. CMB dipole in a map of the whole sky in a projection of Mollweide, as seen by the WMAP satellite. The colors (purple, blue, green, yellow, orange, red) represent the anomaly from lower to higher temperatures. Credit: NASA/WMAP Science Team.

A completely homogeneous and isotropic radiation would, however, have implied the impossibility of the formation of cosmic structures and therefore also of life. For this reason cosmologists began to suspect that there must have been inhomogeneities. They were observed in 1992 by the COBE satellite in an experiment conducted by George Smoot and John Mather, who, for the discovery, won the Nobel Prize in 2006. The observations showed that the background radiation is isotropic up to 1 part in 100,000.

In 1990 Mather, using COBE, also found that the radiation had the emission of a perfect *black body*, that is, of an ideal object that absorbs all the incident radiation without reflecting it, with a temperature of 2.725 K (Fig. 2.5). The observation confirmed Richard Tolman's prediction, in 1934, that the blackbody radiation in an expanding universe cools, but retains the same shape (continues to be described by a blackbody distribution at different temperatures). So the shape of the spectrum of the CMB, dating back to two months after the Big Bang, remained unchanged to the present day.

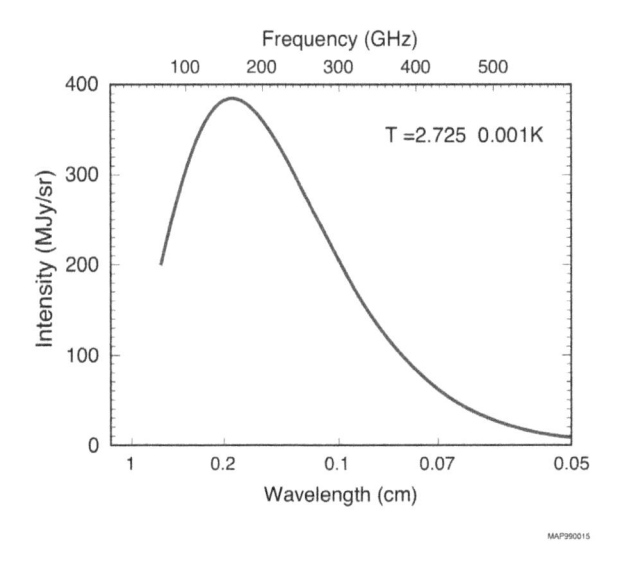

Fig. 2.5. CMB blackbody spectrum. The line represents the theoretical predictions of blackbody radiation. The experimental data perfectly agree with this curve and are not plotted. Credit: NASA/WMAP Science Team.

Given the great importance of the CMB, experiments were carried out to measure its characteristics with an increasing degree of precision. This was done with subsequent experiments on balloons, such as BOOMERANG, or satellites such as WMAP and PLANCK, which allowed an even more in-depth study of the inhomogeneities, dubbed *density fluctuations*, of the background radiation, due to differences in temperature. Starting from 1997, three flights of a high-altitude balloon, the BOOMERANG experiment (Balloon Observations Of Millimetric Extragalactic Radiation and Geophysics), were carried out. In 1997 the balloon flew to the skies of North America and in 1998 and 2003 to the skies of Antarctica. Like a real boomerang, taking advantage of the polar vortex, the balloon departed from the McMurdo base, flew to an altitude of 42 km, to reduce the absorption of the CMB microwaves by the Earth's atmosphere, and with a closed orbit returned to the point of departure. The experiment, directed by Paolo de Bernardis and Andrew Lange, provided a high-resolution image of the CMB anisotropies in a portion of the sky. Given the success of the experiment, NASA designed a space mission, WMAP (Wilkinson Microwave Anisotropy Probe), launched in 2001, followed by the PLANCK mission of the ESA (European Space Agency), launched in 2009.

The satellites produced a map of the background radiation consisting of orange zones, warmer than average, and blue, colder than average, highlighting the distribution of the plasma temperature of which the universe was made up 380,000 years after the Big Bang (Fig. 2.6). The map represents the small temperature variations coming from the surface where the photons interacted for the last time with an electron, the so-called *surface of last scattering* (Fig. 2.7), a sphere of the primordial universe in the center of which we are located, or rather the surface of the cloud from which the light was scattered for the last time. Since recombination was not an instantaneous process, one must imagine this surface as a spherical shell with a thickness of about 30,000 light-years.

Fluctuations are isotropic up to 1 part in 100,000, and in the map (Fig. 2.6) they are visible because they are

The Primordial Universe

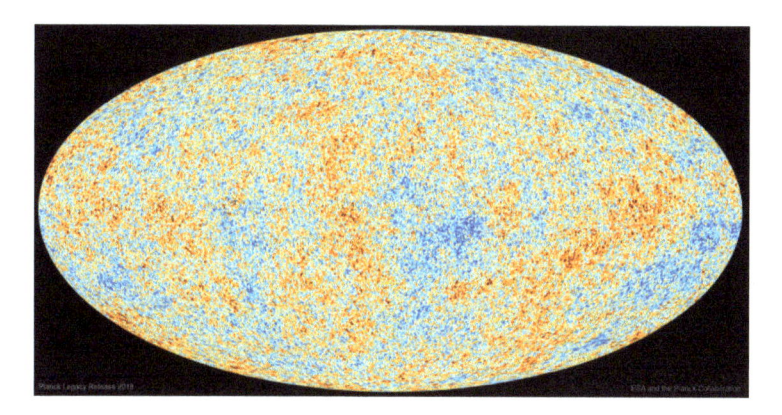

Fig. 2.6. Map of CMB anisotropies obtained by the PLANCK mission. Credit: ESA and the Planck Collaboration.

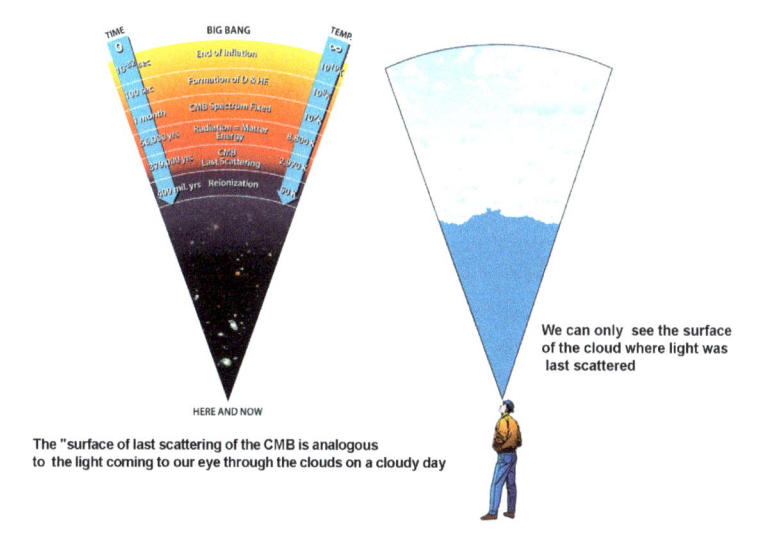

Fig. 2.7. Analogy between observation on a cloudy day and the limits of observation in the primordial universe. On a cloudy day you cannot see beyond the clouds. Similarly in the universe we cannot see beyond the surface of the last scattering. Credit: NASA/WMAP Science Team.

greatly amplified. The hottest areas in the map correspond to the densest regions, so the map also represents *density fluctuations* in the early universe, and could be defined as a photo of the universe at the time of recombination. The scales and magnitudes of these fluctuations determine what our current universe is. Fluctuations are loaded with information about the primordial universe, which we will discuss in Chapter 4.

2.4 The Dark Ages of the Universe

After recombination the universe was made up of hydrogen, helium, small quantities of deuterium, and hydrogen molecules (H_2), formed during the era of *nucleosynthesis*. The photons of the CMB due to the expansion of the universe lost energy, and the wavelengths were moved up to the infrared, causing the universe to be deprived of visible light. Since there were no light sources, like the stars, the universe was transparent, but dark. This period is called the *dark era*, and developed in the period between 380,000 years and the formation of the first stars, that is, a few hundred million years after the Big Bang.

The only source of photons was the hydrogen atoms that emitted 21 cm radio waves, the so-called *HI line*, first observed in 1951 by Harold Ewen and Edward Purcell. This line is of fundamental importance to map neutral hydrogen in space. Hydrogen atoms are stable, and isolated neutral hydrogen usually does not emit radiation. When the spin of the proton and that of the electron are aligned, the hydrogen atom is in a state of higher energy than when they are misaligned. When passing from a state in which the spins are aligned to one with spin misaligned, the neutral hydrogen emits a particular photon with a wavelength of 21 cm, which indicates where the atom is located (Fig. 2.8).

These photons could theoretically be observed by radio telescopes, but so far this has not happened. Their detection would clarify how matter was distributed in the dark age, how primordial stars formed, and the role of recombination in the formation of galaxies.

The duration of this era is not exactly known, because it is not known with certainty when the first stars formed. How may we estimate the time when they started to form?

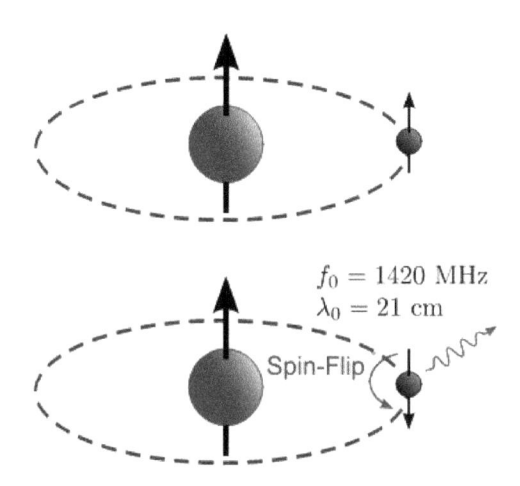

Fig. 2.8. Generation of the 21 cm line of neutral hydrogen. The line comes from the electron transition between two levels of the hydrogen ground state. Credit: Wikipedia.

To this aim, the *Gunn–Peterson effect* is used. It detects the time at which the universe was re-ionized by those stars.

In 1965 James Gunn and Bruce Peterson proposed to use quasars (very old, distant, and very bright galaxies) to date the final stages of re-ionization. According to the two researchers, neutral hydrogen would have produced a significant depression in the high redshift quasar spectrum, at wavelengths shorter than 121.6 nanometers.[8]

That is, if the quasar is so far away that its radiation was emitted in the *dark era*, the neutral hydrogen present at that time will absorb its ultraviolet (UV) radiation, while this will not occur if the quasar is closer (Fig. 2.9). Figure 2.9 shows two quasars at different redshifts, that is, at different distances and epochs. It reveals that a quasar at $z = 6.28$ (888 million years after the Big Bang) (bottom panel) shows the absorption, while the one with $z = 5.99$ (944 million years after the Big Bang) (top panel), closer to us than the other, does not show it.

---

[8] A nanometer equals one billionth of a meter.

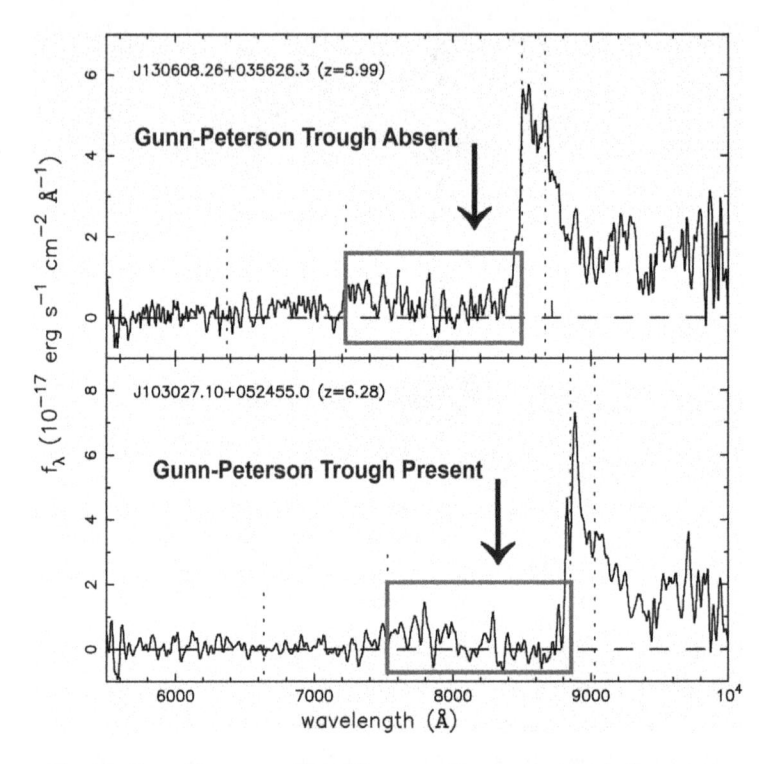

Fig. 2.9. Gunn–Peterson effect. The quasar in the top panel, closest to us, shows no absorption, while the quasar in the bottom panel further away shows absorption. Credit: Modification of work by Xiaohui Fan et al. (*The Astrophysical Journal*, 2006, 132, 117).

In 2001, a group of scientists led by Robert Becker confirmed the Gunn–Peterson effect using the Sloan Digital Sky Survey, which is an extensive astronomical survey. The observed absorption is found in the infrared band of the spectrum, due to the huge distance of the quasar. Another method used to date recombination is based on the CMB. The WMAP satellite, in a study carried out between 2001 and 2006, established that re-ionization began about 400 million years after the Big Bang and ended after 400–500 million years, according to quasar studies: 900 millions of years after the Big Bang.

So the dark era started after recombination, 380,000 years, and ended about 300 million years after the Big Bang, when

The Primordial Universe

the first stars and galaxies were formed, which reionized the universe by plunging it back into the dark, until about a billion years after the Big Bang. From some studies of the PLANCK satellite, the first stars were born 300–400 million years after the Big Bang, but most were born around 700 million years after the Big Bang, a little later than previously thought.

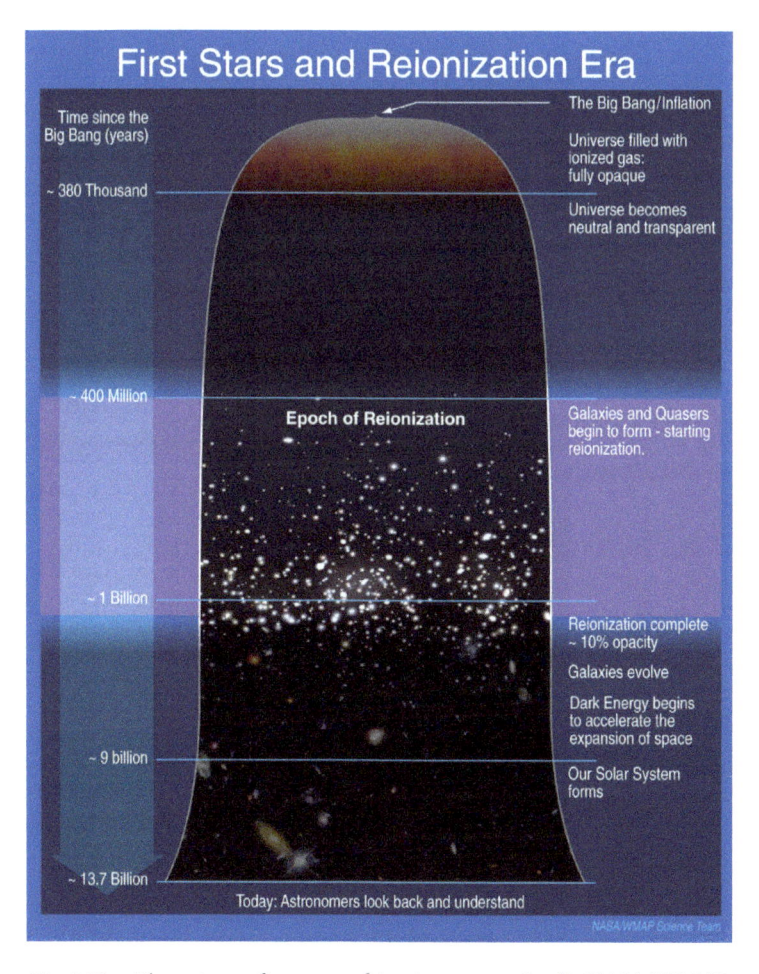

Fig. 2.10. The universe from recombination to now. Credit: NASA/WMAP Science Team.

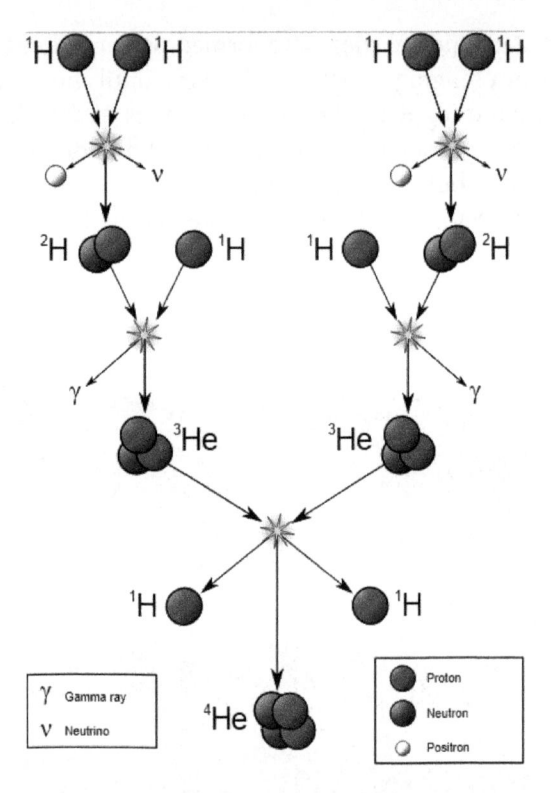

Fig. 2.11. Proton–proton chain: 4 protons form a helium nucleus, ⁴He. Credit: Wikipedia.

2.5 The First Stars

In the dark era, the universe was dark until it was brightened by the first stars. These stars, not yet observed, were different from the present ones. In the first place they were much larger than today's stars: a few hundred times larger than the Sun. The mechanism that produced energy was, as in today's stars, nuclear fusion, and in particular the dominant mechanism in the Sun: the *proton–proton chain*, in which four hydrogen nuclei are transformed into one of helium. As seen in Fig. 2.11, two protons come together to form *deuterium*, a hydrogen *isotope.*[9] To these is added a proton forming *tritium* with release of energy. Finally, the union of two tritium isotopes forms helium made of two protons and two neutrons. The mass of the helium nucleus is less than the sum of the masses of the reacting nuclei, the four protons. This mass difference, 0.7% of the original mass, for the mass energy equality discovered by Einstein, is released in the form of energy, equal to 26.73 MeV. This energy is equal to that produced by burning approximately 11 tons of coal.

In the stars more massive than the Sun, there is another type of process in which, again, four protons go to form a

---

[9] Isotopes are atoms having an equal number of protons but different number of neutrons. For example, the hydrogen nucleus has only one proton, while its isotope called deuterium contains a proton and a neutron, and tritium two neutrons and a proton.

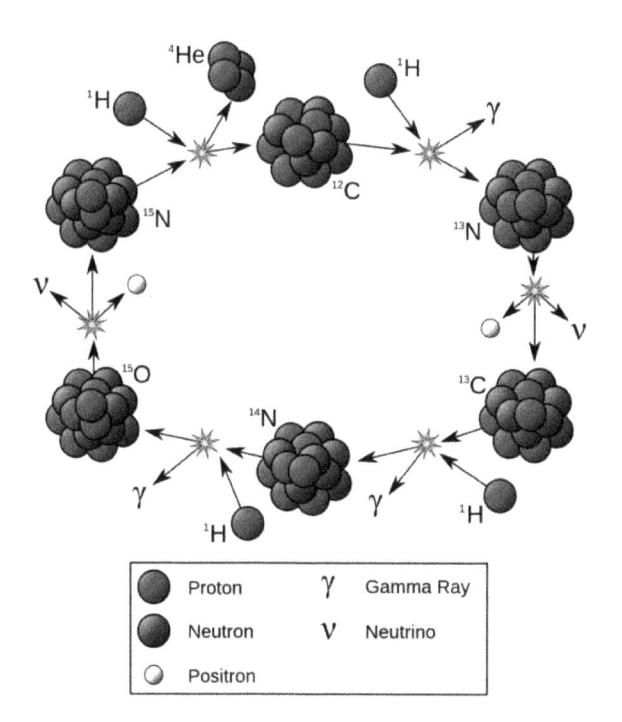

Fig. 2.12. Carbon cycle. Formation of helium, ^{4}He, through the "help" of carbon, nitrogen, and oxygen. Credit: Wikipedia.

helium atom, with the help of carbon, nitrogen, and oxygen (Fig. 2.12), dubbed the *carbon cycle*. Because of the absence of heavy elements such as carbon, the *carbon cycle* could not function in primordial stars. Furthermore, if the first stars had worked with this mechanism, nuclear processes would have been very fast and their life would have been very short. With the proton–proton chain, their life is a little longer, about 10 million years, in any case a very short life compared to the life of a star like the Sun, which lives about 10 billion years.

Furthermore, these stars, having formed from the gas that filled the universe, namely hydrogen and helium, were made up only of these elements. During their lifetime they produced elements heavier than hydrogen, and when they exploded as a *couple-instability supernovae*,[10] they scattered them in space. The daughters of these stars were formed from the remnants of primordial stars and contained traces of heavier elements besides hydrogen and helium. The primordial stars, called *population III stars*, in a hundred million years released a considerable amount of ultraviolet (UV) radiation which ionized the gas. The universe became opaque again. The expansion of the universe diluted the density of the plasma, until the photons were again free to move in the universe. So the universe passed through more than one *fiat lux*: with the Big Bang, after recombination, and finally at the end of re-ionization.

2.6 The First Galaxies and the Structure of the Universe

After the formation of the stars, structures such as galaxies formed. We will cover some details of the topic in Appendix B. Here we just want to give an idea of the formation of galaxies and structures in general, summarized in Fig. 2.13.

In the dark ages that we are considering, as in the rest of cosmic history, the expanding universe contained a gas made

---

[10] They are supernovae due to the production of electron–positron pairs.

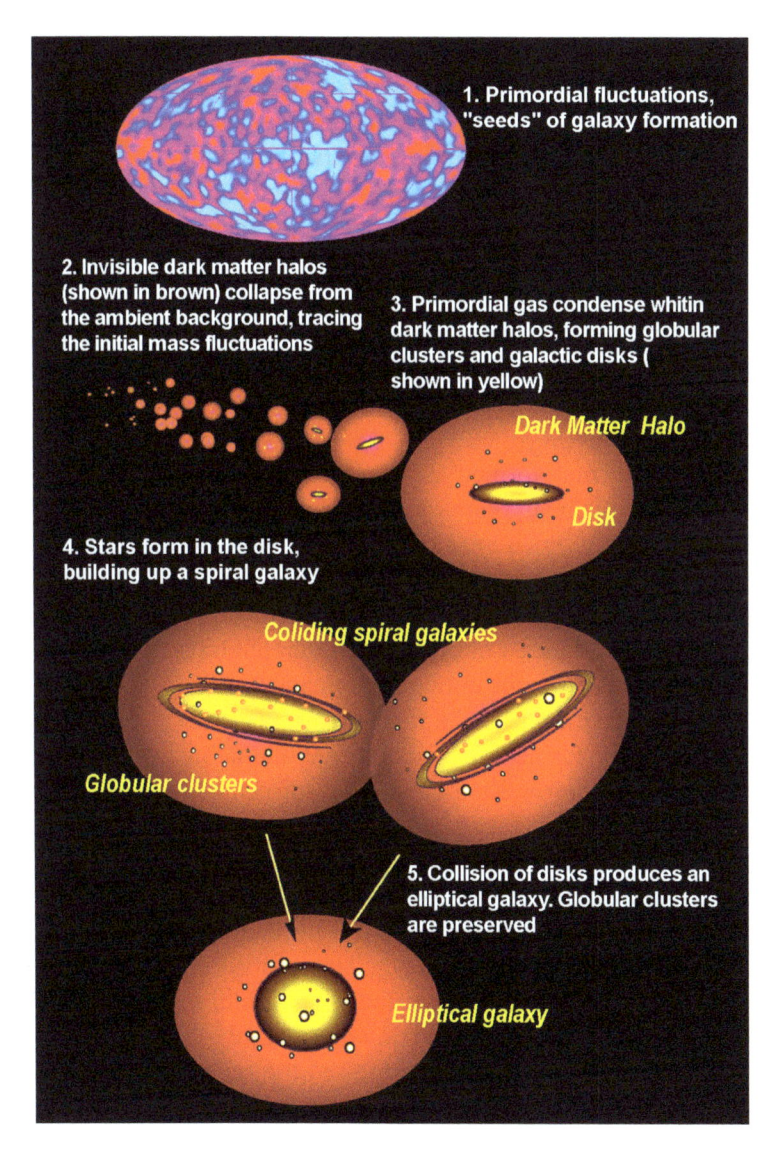

Fig. 2.13. Scheme of formation of spiral and elliptical galaxies starting from fluctuations in primordial density. Credit: Modification of work by Roberto G. Abraham and Sidney van den Bergh (*Science*, 2001, 293, 1273–1278).

of light elements and dark matter much more abundant than gas. What happened to the gas and dark matter present in the universe? Denser areas could be found in the primordial universe. These areas are the seeds from which the structures in the universe were formed. The large mass content of the quoted regions generated a larger gravitational force with respect to neighboring regions, thus overcoming the expansion of the universe. Finally, those mass concentrations, mainly constituted by dark matter, could collapse to form the so-called *dark matter halos*. The first that formed according to some numerical simulations could have masses much smaller than the Sun, and recent observations from the Hubble telescope found clumps 10^{-5} times our galaxy's dark halo. Gravity helped them to grow, aggregating matter from the surroundings, so much so that today the largest of these halos have masses of 1 million billion Suns ($10^{15}\ M_{\odot}$).

Then, dark matter first formed dark structures, halos, which, from the gravitational point of view, behaved like some sort of wells within which later the gas fell, giving rise to visible objects such as galaxies. Dark matter formed its invisible structures before baryons because it was not subject to the action of photons. In the case of ordinary matter, photons did not allow the atoms to form until recombination. After recombination, ordinary matter, in the form of gas, began to be attracted to the "wells" formed by dark matter, illuminating them. Thus galaxies were born. The first galaxies formed had irregular shapes. By aggregation they gave rise to spiral galaxies. The collision of spiral galaxies gave birth to elliptical galaxies.

The formation and evolution of galaxies are well illustrated in the image, called the *Hubble Deep Field* (HDF), containing about 10,000 galaxies and obtained using 800 exposures. It allows us to look at the universe from today to 13 billion years ago. In 2012, another image of a portion of the HDF was published combining 10 years of photographs of the region obtained from the Hubble Space Telescope. This image, called *eXtreme Deep Field* (XDF), shown in Fig. 2.14, is the deepest image of the sky ever obtained that allows us to go back in time for 13.2 billion years. The image shows the dramatic emergence of violently growing galaxies through collisions and mergers. Blurry, fuzzy red galaxies are observed, the remnants of dramatic collisions between galaxies. Scattered in the image, tiny and weak galaxies are observed, more

The Primordial Universe

Fig. 2.14. Image of the XDF. Credit: NASA, ESA, G. Illingworth, D. Magee, and P. Oesch (University of California, Santa Cruz), R. Bouwens (Leiden University), and the HUDF09 Team.

distant than the red ones, which constitute the seeds from which today's galaxies were born. The youngest galaxies in the XDF formed just 450 million years after the Big Bang. The image (Fig. 2.14) is a confirmation of the *hierarchical model* of formation of structures in the universe. Structure formation starts from smaller objects and, by aggregation, forms larger objects through their merger.

Spiral galaxies are made up of young stars (millions of years) and gas structured in a disk with spiral arms, a spherical central area called the *bulge*, which contains older stars (billions of years). The size of the stellar distribution, in an average spiral, is of the order of tens of kiloparsecs. The average elliptical galaxies have little gas and have a spheroidal shape, with stellar masses of the order of 10^{12} solar masses, a dozen times greater than the spiral ones. There are also *dwarf spheroidal* and *irregular galaxies*, the shapes of which are not well defined. Galaxies tend to cluster. Our galaxy is located in the local group, having a size of 1–2 Mpc, consisting of

The Invisible Universe

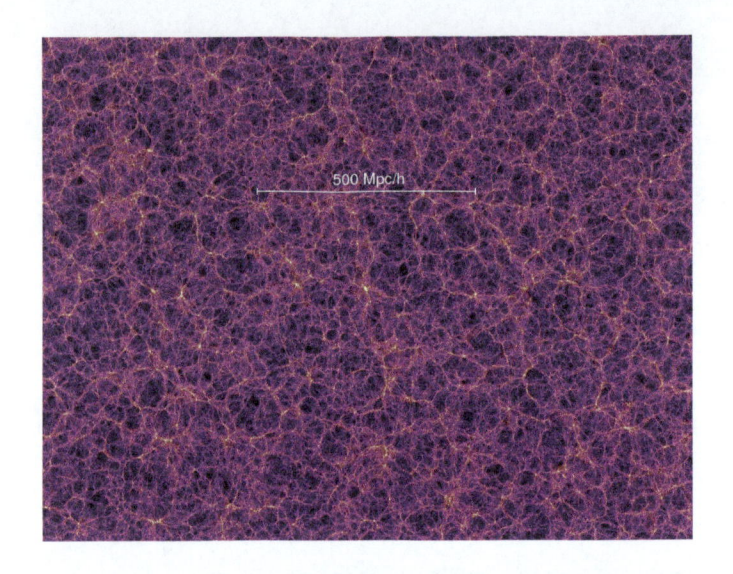

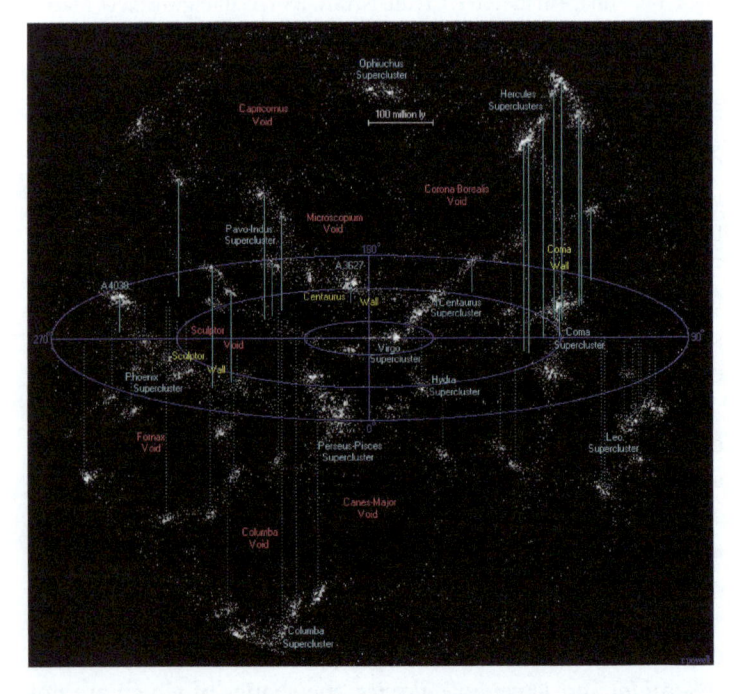

Fig. 2.15. Top: Simulation of the large-scale structure of the universe. Credit: Millennium Simulation Project. Bottom: Distribution of superclusters and voids in the universe. Credit: R. Powell (http://www.atlasoftheuniverse.com/nearsc.html).

3 main galaxies: ours, Andromeda, and the Triangle, and about 70 dwarf galaxies. Large groups of galaxies are called *clusters of galaxies*, which contain thousands of galaxies, with dimensions of about 10 megaparsecs and 10^{14}–10^{15} solar masses.

On a still larger scale can be found *superclusters* with sizes of a hundred megaparsecs. There are also spherical regions with very few galaxies called *voids*, 10–100 Mpc. The large-scale structure of the universe consists of *filaments*, and *walls*. Clusters form at the crossing points of the filaments. The whole thing resembles a spider's *cosmic web*, as seen in the computer simulation shown in the top panel of Fig. 2.15. The bottom panel of Fig. 2.15 shows the universe over hundreds of millions of light-years. All the groups of white dots with names in blue are superclusters, and the dark areas with red writing are the *voids*. We are located in the *Virgo cluster* in the center of the map. The *large-scale structure* of our universe resembles that of a sponge.

Chapter 3

HOW DO WE KNOW THAT DARK MATTER EXISTS?

What is essential is invisible to the eye.

— Antoine de Saint-Exupéry

3.1 Astronomy of the Invisible

Sirius is the brightest star in the sky. Its motion does not follow a straight line, but writhes in a serpentine manner. This betrays, in accordance with Newton's laws, the presence of a companion not easily visible, at least with the technology of the times of Friedrich Wilhelm Bessel, a great German mathematician and astronomer. Bessel came to astronomy following tortuous paths such as those followed by Sirius. Given his difficulties in Latin (which he later curiously learned so well that he was able to teach it), he employed himself as an accounting apprentice, while studying geography at night. He became a cargo officer and began to take an interest in determining the position of a ship on the high seas, thus leading himself to astronomy. He distinguished himself for determining the distance of the star 61 Cygni, calculated the orbit of Halley's comet and realized, in 1844, from the study of the motion of Procyon and Sirius, that these stars had invisible companions. This study was baptized by him as *astronomy of the invisible*, a very current definition, as the astrophysics research of dark matter is nothing but astronomy of the invisible. The companions of Sirius and Procyon were observed a few years later: the first in 1862 by Alvan Graham Clark and his father, Alvan Clark, and the second in 1896 by John Schaeberle. However, success is not always connected to skill, to solid models and interpretations as in Bessel's case, but also to chance. An example is the discovery of Pluto. The motion of Uranus and Neptune seemed to be perturbed by some external object, which was called planet X by Percival Lowell. In 1830 Clyde Tombaugh discovered the planet X, called Pluto, in almost the same position predicted by the calculations. This was only a coincidence because, as we know today, Pluto does not give rise to perturbations in the motion of Uranus and Neptune, given its small size, determined in 1989 by Voyager 2. The history of these errors related to the astronomy of the invisible continued with Mercury. Le Verrier noticed orbital anomalies, a precession of the perihelion of Mercury of 574 arc seconds per century[1] (Fig. 3.1). After subtracting 531 arc seconds due to perturbations generated by the other planets, 43 arc seconds remained. Le Verrier considered them as evidence of a planet orbiting the Sun and Mercury, and that planet was called *Vulcan*. The absurd thing is that, in 1878, James Watson and Louis Swift announced the observation of Vulcan, which, however, as it does not exist,

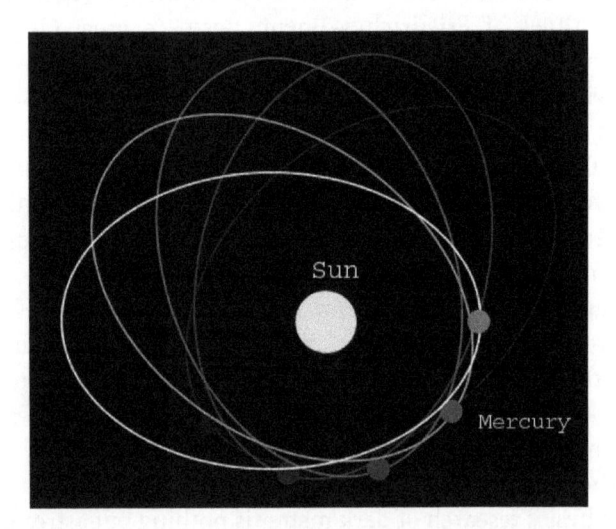

Fig. 3.1. Precession of the perihelion of Mercury. Credit: Wikipedia, Rainer Zenz.

---

[1] Mercury's perihelion (maximum distance from the Sun) is not fixed in space, but precedes; that is, it changes position.

was not confirmed by other observations. The problem was solved when Einstein applied general relativity, published in 1915, to the orbit of Mercury.

It reproduces exactly the 43 seconds of arc missing from the Newtonian prediction. The other planets in the Solar System also have small precessions explained by general relativity.[2]

This discrepancy between theory and observations highlights a fundamental point: an anomalous motion of an object such as a planet or an entire galaxy can be explained in two ways:

- with the presence of invisible masses, that is, with the presence of *dark matter*
- by changing the gravity theory

This is still valid today. There are supporters of the existence of dark matter and scientists who think that it does not exist, and that simply the theory of gravity that we use is incorrect. This prompted to propose a whole series of *theories of modified gravity*.

Other objects that can be counted in the astronomy of the invisible are *black holes*. John Michell was the first to think about objects so massive that light cannot escape, and the term *black hole* was introduced a few centuries later by John Archibald Wheeler.

Another idea about the nature of dark matter is that it was made up of dark nebulae. One of the first to discuss the topic, in the second half of the 19th century, was Father Angelo Secchi. The observation of dark areas in fields densely populated by stars was interpreted in different ways: as fields with few stars or as masses that absorbed light.

---

[2] Among the many oddities that fate put on the table in relation to the confirmation of general relativity, there is one very peculiar. In the first unsuccessful attempt to verify the theory in the 1914 eclipse in Crimea, an expedition led by the astronomer Freundlich (friend of Einstein) made the acquaintance of Argentine astronomers who had gone to Crimea to observe the same eclipse, just to reveal the existence of Vulcan.

The problem of *missing matter* can be considered as the problem of unobserved stars and planets. By observing astrophysical systems such as galaxies or clusters of galaxies, we observe anomalies in the speeds of stars in galaxies or galaxies in clusters, too high compared to what the mass of the object allows. These anomalies are explained by a greater content of invisible material in the universe or by assuming that our knowledge of the laws of gravitation are incorrect.

3.2 First Weak Evidences

The stars of our galaxy located on its disk move in approximately circular orbits. In reality the observed motion is composed of three oscillating motions in a direction perpendicular to the plane and in those orthogonal to it. Looking at the galaxy edge-on, the stars will move up and down like a rocking horse.

It is obvious that a greater quantity of local mass will produce a greater attraction force on the stars and therefore a lower amplitude in the stellar oscillations. So the study of stellar dynamics allows to establish the local mass content. One of the first astronomers to carry out the problem of determining local density was Lord Kelvin, assuming that the stars behaved like particles in a gas. Poincaré, from Kelvin's results, concluded that the amount of dark matter (term introduced by him: "matiere obscure") had to be less or of the same order as the ordinary one. Jacobus Cornelius Kapteyn used his model of the Milky Way, now known as the Kapteyn Universe, together with the vertical motion of the stars in the galaxy, concluding that the amount of dark matter had to be low. Another astronomer, Öpik, had come to similar conclusions, while a re-analysis of Kapteyn's results by James Jeans led to different conclusions: stellar motion required a mass greater than that observed. These conclusions were confirmed by a student of Kapteyn, Jan Oort. He observed the vertical motions of the stars noting that the motions were faster than what can be expected assuming that the mass had a distribution equal to that of the visible mass. He established a value for the total mass near the Sun and determined a limit for dark matter in the galactic disk, known as the *Oort limit*. The study of the amount of dark matter in the galactic disc is still a problem of current importance because this value is necessary in dark matter detection experiments.

3.3 How Much Does a Cluster of Galaxies Weigh?

A fundamental step on the road to the study of dark matter is the one Fritz Zwicky took in the 1930s. Zwicky was born in Varna to a Swiss father, and using an exaggeration, one could say that he is perhaps more known for his grumpy character than for his scientific discoveries, although of fundamental importance. Together with Baade he coined the term *supernova*, and he hypothesized that neutron stars are formed during the collapse of these stars. Zwicky also hypothesized how *cosmic rays* could be produced by supernova explosion, proposed using objects more massive than stars to verify the *gravitational lens effect* (we will talk about it later) predicted by general relativity.

As for dark matter, in works from 1933 and 1937, Zwicky studied the intrinsic velocities of the galaxies in the *Coma cluster* and used the velocity to track the amount of mass that made up the cluster.

The structure of the cluster is given by the balance between the energy linked to the motion of the galaxies, that is, the kinetic energy, and the gravitational force. If the kinetic energy is greater than that of gravity, the galaxies run away from the cluster, while if the gravitational energy is greater than the kinetic energy, the galaxies fall towards the center of the cluster. A stable cluster requires a balance between kinetic and gravitational energy, technically dubbed the *virial theorem*.

From the study of the motion of the galaxies in the cluster, Zwicky determined their mass by finding from the virial theorem much higher values than expected. He concluded that in order for the galaxies not to run away from the cluster, a quantity of mass hundreds of times larger than that observed was necessary. He designated this invisible matter with the term *dunkle materie*, from the German for *dark matter*. The result was ignored for 40 years, and it was not the only one of his results that was ignored. In a 1937 article, Zwicky used the Hubble and Humason estimate for the Hubble constant, equal to 558 km/s/Mpc (much greater than the currently used value of ~67 km/s/Mpc, estimated with the space mission PLANCK and ~74 km/s/Mpc using the Hubble Space Telescope). This produces an overestimation of the dark matter in the cluster, which, however, does not affect Zwicky's result.

This method was applied to a large number of clusters, confirming the previous results. As already mentioned, because of his grumpiness, his theories were opposed by colleagues whom he indicated with the nickname "spherical bastards," that is, bastards from any point you observe them.

A few years later, Smith, studying the Virgo cluster, confirmed Zwicky's result and assumed that the missing mass was made up of "internebular matter." The following years were characterized by a long debate between supporters and detractors of the existence of dark matter. There was thought to be another less exotic explanation to what Zwicky deduced. Furthermore, in the case of the Coma cluster, it was not known with certainty whether the cluster was old enough to be in balance to be able to apply the virial theorem.

3.4 Andromeda and Dark Matter

An argument in favor of the existence of dark matter, and different from the previous ones, was that proposed by Kahn and Woltjer. Using observations of the line at 21 cm, it was discovered that Andromeda and our galaxy, unlike the other galaxies, approach one another, with a speed of 125 km/s. Since the two galaxies are part of a gravitationally linked system, the two determined the mass of the system[3] with Keplerian dynamics, which was about 20 times greater

---

[3] The mass can be determined if the distance and approach speed are known, assuming an orbital period lower than the age of the universe and a radial orbit.

than the sum of the masses of the stars of the two galaxies: $M > 2 \times 10^{12}\ M_{\odot}$. According to current knowledge, this argument clearly showed that the galaxies were immersed in a halo of dark matter. Kahn and Woltjer thought that the mass excess was made up of ionized intergalactic gas.

Although the argument was correct, straightforward, and simple, Kahn and Woltjer's article did not have much impact, similarly to Zwiky's articles of 1933 and 1937, probably due to the absence of a framework in which to interpret the observations of a total mass so great for the Andromeda–Milky Way system.

3.5 The Rotation of Galaxies

The local mass distribution of a spiral galaxy can be determined by using the vertical motions of the stars, and also by using the circular motions of the stars on the galactic disk. Like the vertical motion of the stars, the circular one is characterized by a speed that depends on the mass inside the stellar orbit. The stellar speed in terms of distance is called *rotation curve*. The study of the circular motions of stars in galaxies was started in 1939 by Babcock, who noticed that the stars in the outer part of Andromeda were moving with unexpectedly high speeds. The growth of the rotation curve, at great distances from the center, led one to think that a large amount of matter existed in the outer parts of the galaxy, but Babcock gave an interpretation that excluded the connection with dark matter.

Before him, Lundmark had obtained, for some galaxies, high values of the mass–brightness ratio, which is an indicator of the presence of mass in excess. Holmberg interpreted these results as due to the absorption of light by dark matter. He erroneously concluded that taking into account the values obtained by Lundmark would have reduced the results to typical values.

Although the studies discussed (Zwicky, Kahn, and Woltjer) clearly showed that the clusters and galaxies contained more mass than the visible one, the situation only became clear in the 1970s. The pioneer of this change was Vera Rubin[4] with her collaborator Kent Ford. The two started an Andromeda study in 1970. Due to the rotational motion of the stars in the disk, their light when approaching the observer undergoes a shift towards blue and that from the stars moving away shows a shift towards red, due to the well-known *Doppler effect* (see Box 3). The change in the wavelength of light is proportional to the speed of the source. In this way it was possible to determine the rotation speeds in different positions of the disk. As in the case of the planetary orbits, Rubin and Ford expected a growth in the speed of rotation up to a maximum, and therefore, moving away from the center, a decrease in speed, the so-called *Keplerian fall*.

What they observed was completely unexpected: the rotation speed of the stars at great distances from the center did not decrease, but was similar to those of the stars closest to the center (Fig. 3.2). The rotation curve had a flat structure.

This was extremely strange because moving outwards the stellar mass decreased, and with it the rotation speed would have to decrease.

The flatness of the rotation curves of the galaxies is strong evidence of the existence of dark matter. If the rotation speed is constant, this implies that the mass must grow with the radius as shown by the solid line in the left panel of Fig. 3.3, beyond the point where the light of the stars is no longer observed. Furthermore, this implies that the galaxy must be contained within a huge region of non-visible matter (Fig. 3.3, right panel) — a great deal of mass that we don't observe.

---

[4] Vera Rubin became fond of astronomy at an early age. She used to look at the stars and the sky from her room facing north to her home in Washington D.C. After finishing college, she was unable to continue her studies at Harvard because women were not accepted at that time, and this until 1975. She continued her studies at Cornell University where she studied with famous physicists such as Richard Feynman and Hans Bethe. She obtained her doctorate from the University of Georgetown under the guidance of another famous physicist, George Gamow. She obtained a research position at the Carnegie Institution in Washington and there she began her observations together with Kent Ford.

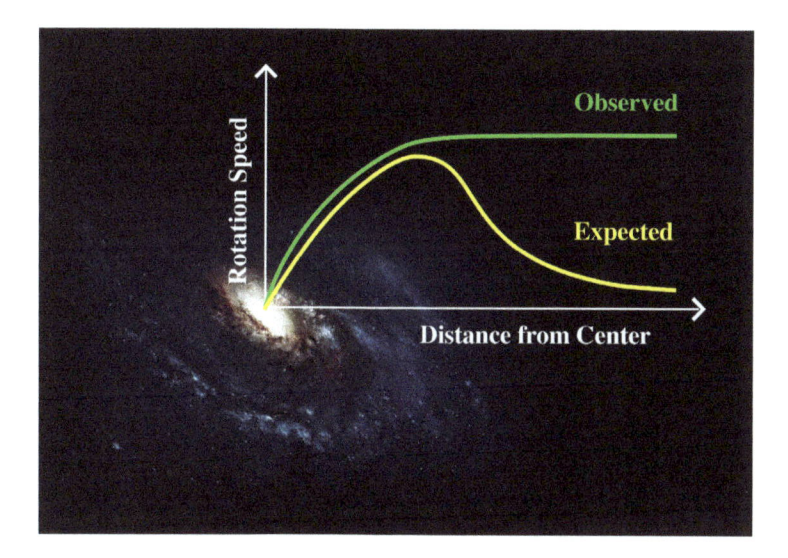

Fig. 3.2. Typical rotation curve of a galaxy.

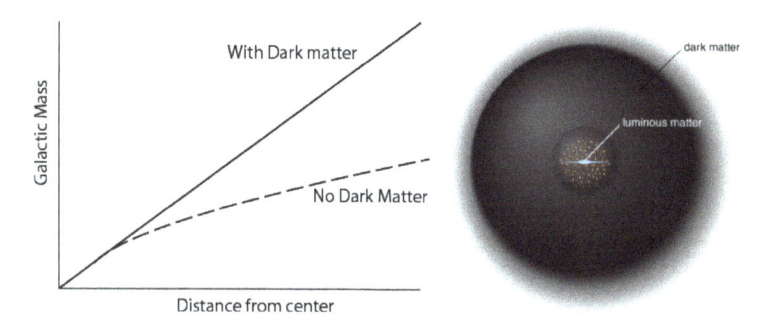

Fig. 3.3. Left: Mass predicted by the rotation curve (solid line). Mass that would be expected in a system made up only of stars and gas (dotted line). Right: Galaxy in the center of a dark matter halo. Credit: Modification of work by Infn/Internosei.

Observations indirectly indicated that a large amount of non-visible mass had to exist. How is this mass distributed? An approximate idea can be obtained by remembering that Newtonian mechanics tells us that a constant rotation speed implies a linear growth of the mass with the distance from the center of the galaxy. That is, the mass must be proportional to the radius (see Fig 3.3, left panel). Since the density is inversely proportional to the cube of the distance and proportional to the mass, and the mass is proportional to the distance, one can conclude that the density is inversely proportional to the square of the distance. This means that the galaxy is immersed in a spherical distribution of dark matter called a dark matter halo (Fig. 3.3, right side). More precise models show that the halo has the shape of an ellipsoid.

Several other authors came to similar conclusions (e.g., Freeman, Roberts). In 1974 two groups led by Einasto and Ostriker declared in two important articles that the mass of the galaxies was underestimated by a factor of 10. The mass excess according to them would have been gas in the outer part of the galaxies. The result related to mass excess was confirmed by other studies. In particular, the 1978 Bosma study showed that using radio band observations, one obtained flat rotational curves at much greater distances than those studied in the optical band. Furthermore, it was seen that the mass continued to grow well beyond the region occupied by stars and gas. Rubin, Ford, and Thonnard could not see the expected fall in the rotation curves, even studying the galaxies up to the maximum observable optical distance. This was so strange and unexpected that, in the following years, there was talk of a *disk-halo conspiracy*. It was as if the disk and the halo had agreed to generate a flat rotation curve.

Rotation curves are an important indirect evidence of the existence of dark matter, and also indicate how it is distributed within galaxies. Ordinary matter is dominant in the inner part of a galaxy, up to a few kpc, and when moving away from the center, dark matter becomes already dominant at about 10 kpc. This trend is more easily revealed in spiral galaxies using neutral hydrogen as a speed tracer, that is, the 21 cm line, and is also similar in galaxy clusters. In general, there is an increase in the amount of dark matter going from galaxies to clusters, whose matter content is made of 95% dark matter.

For the sake of completeness, we want to recall another evidence in favor of the existence of dark matter that comes from a 1973 study by Peebles and Ostriker, who, by means of pioneering simulations, showed that to stabilize the structure of the disks of the spiral galaxies, the presence of massive halos around the galaxies is necessary.

3.6 Radiography of Galaxies

Other evidence in favor of the existence of dark matter came from the detection of emission in the X-ray band from the very hot gas (10–100 million degrees) present in elliptical galaxies and clusters. The Virgo cluster located about 60 million light years away from us, contains a thousand galaxies, including, in its center, a giant elliptical galaxy, M87. Radiation in the X-ray band was observed in 1966 from this galaxy, and a few years later the same type of radiation was observed in the Coma (Fig. 3.4) and Perseus clusters. The observed gas is in hydrostatic equilibrium, that is, the particles that make up the cluster move in the gravitational field of the cluster with speeds corresponding to the attraction of its mass.

The mass distribution in the clusters can be determined if the temperature and density of the gas are known. These

Fig. 3.4. X-ray emission of the Coma cluster.

Source: Image from the Chandra telescope. Credit: X-ray: NASA/CXC/University of Chicago, I. Zhuravleva *et al*. Optical: SDSS.

quantities can be determined using telescopes that observe the X-rays emitted by the cluster. This was done by the Einstein satellite (and later by other satellites). In clusters, the gas is trapped by gravity. However, alone, the gravity generated by the stars is not sufficient to trap the gas. In order to avoid the gas from flying away, a huge amount of invisible mass is needed: dark matter. The results confirmed previous estimates made by Zwiky and other astronomers on the basis of the virial theorem. The results of the hot gas observations led to the conclusion that about 5% of the mass was made up of galaxies, 15% of gas, and 80% of non-visible matter. This matter is not interstellar material, which despite not being visible in the optical band can be seen with infrared-sensitive telescopes, but matter that is not seen directly by any kind of telescope.

3.7 The Tricks of Light

Light is a wave produced by the oscillations of the electromagnetic field. In empty space, and in the absence of masses, light moves in straight paths, called rays. As we know from daily experience, observing a teaspoon immersed in a glass of water, the object seems to be broken into two pieces: a part above the water and a part below. The deviation of the rays in the passage from air to water, or in general between two different media, is known as *refraction*. According to *Fermat's principle*, obtained by a well-known 17th-century mathematician, in the motion between two points, light follows the path of minimum time. Using his principle, Fermat was able to determine the law of refraction, known as the *Snell–Descartes law*. Since light moves faster in the air than in water, Fermat's theorem implies that the trajectory is not straight, and that the ray out of the water has a different inclination than that in the water. The difference in inclination between the two rays depends on the refractive index, that is, the ratio between the speed of light in empty space and that in a medium. The behavior of light is similar to that of a lifeguard who tries to save a swimmer in difficulty. Knowing that he can move faster on land than at sea, he will choose a longer trajectory on the beach in order to minimize time, following a *brachistocrone curve*, that is, the minimum time path (the red curve in Fig. 3.5).

Similarly, if light has to travel between two points between which the temperature, and therefore the density, are different, it will move in a

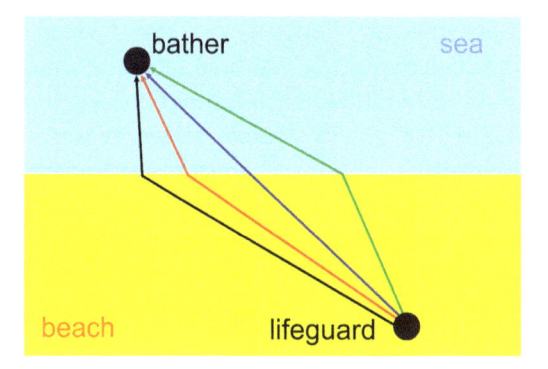

Fig. 3.5. The curve in red is the trajectory followed by a lifeguard to save a bather.

curved path. What happens in the presence of a medium, such as water or air, can also happen in empty space, in the presence of masses, as Einstein showed.

3.8 Cosmic Mirages

In Newtonian mechanics time and space are two absolute quantities, rigid and immutable. Starting in 1907, by trial and error, with work of almost eight years, Einstein obtained his gravitational field equations and presented them, in November 1915, at the Prussian Academy of Sciences. The starting point of the path bringing Einstein to formulate the theory was the *principle of equivalence*, which Einstein remembers as the happiest idea of his life. It states that around any point it is always possible to find a reference system in which the effects of acceleration due to the gravitational field are zero. In Einstein's words:

> When I was busy (in 1907) writing a summary of my work on the theory of special relativity ... I got the happiest thought of my life ... Because for an observer in free-fall from the roof of a house there is during the fall, at least in his immediate vicinity, no gravitational field.

This story told by Einstein was transformed into an anecdote in which the falling man was a painter, who, in some versions, told Einstein that he felt himself floating in the air.

The absence of gravity typical of empty space can be recreated by airplanes. In Fig. 3.6, the trajectory of the airplane A300 of the European Space Agency (ESA) is shown. Moving on parabolic orbits, the plane manages to eliminate gravity for parts of its trajectory.

Einstein used this principle as a guide in his search for field equations.

One of the revolutionary consequences of the theory — "the most surprising combination of philosophical penetration, physical intuition and mathematical ability," in the words of Max Born, and "the most beautiful of scientific theories" in those of Lev Landau — is that space and time are not two separate and immutable entities, as Newton thought, but unite in a marriage. They form the four-dimensional space-time, consisting of three spatial coordinates and a temporal one, which behaves like a plastic structure. Both space and time can contract, expand, lengthen, or shorten. This all depends on the mass (and also velocity). The larger the gravitational potential, the slower the time flows.

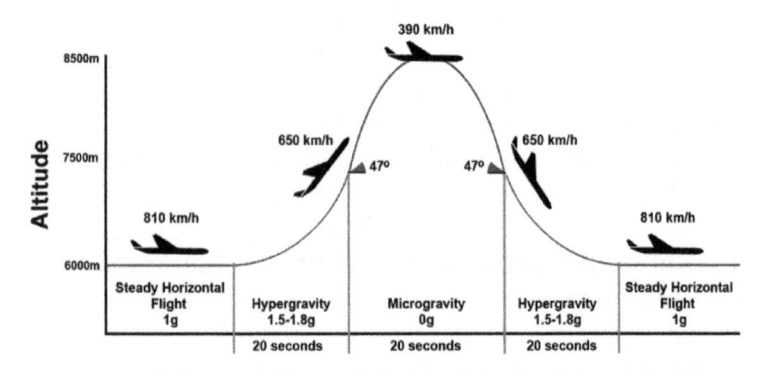

Fig. 3.6. Trajectory of an aircraft (ESA A300) capable of simulating the absence of gravity. In the first part of the trajectory, on the left, the airplane moves horizontally and is subject to the usual gravity. It then moves upwards, accelerating, and is subject to an acceleration 1.5–1.8 times larger than that on earth. In the central part of the trajectory, for 20 seconds, the plane is subject to zero gravity. Credit: Luigi Pizzimenti (https://www.altrimondi.org/il-volo-a-zero-g-dellesa/).

For a person who lives at sea level, time flows more slowly than it does for another person who lives on Mount Everest, since the gravitational field is stronger when we approach the center of the Earth. The time difference, considering an entire human life, is a few billionths of a second. The difference becomes dramatic if we observe an astronaut who approaches the *event horizon* of a black hole, that is, the point of no return. When the astronaut touches the horizon of events, from our point of view, time is frozen. Space shows the same elasticity. In a well-known analogy, space-time is described as a stretched rubber sheet. An object placed on the sheet will deform it, producing a depression (Fig. 3.7).

In theory, light rays, like material bodies, move on trajectories, called *geodesics*, which coincide with the straight line only in empty space.

In general relativity, the motion of the Moon around the Earth is no longer caused by an instantaneous attractive force, an action at a distance between the two objects, as imagined by Newton, but is generated by the deformation of space-time caused by the Earth (Fig. 3.8).

The analogy is not perfect. Firstly, the membrane has only two dimensions, while space-time has four and space three. The whole space around the Earth is deformed. Figure 3.8 only shows a part of the deformation, the deformation under the Earth. Leaving the analogy, the best way to express what

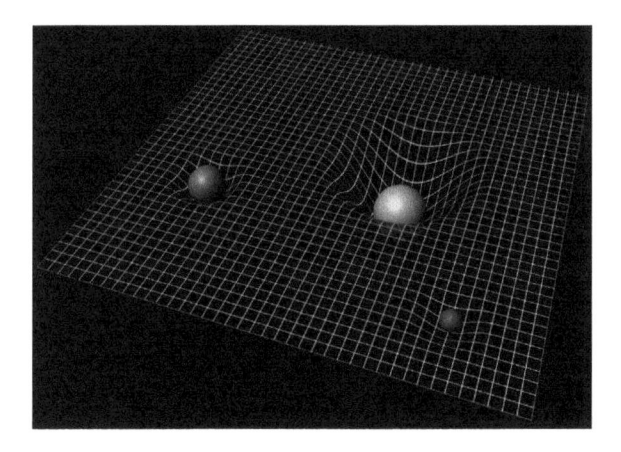

Fig. 3.7. Masses that deform space-time.
Source: ESA-C.Carreau.

Fig. 3.8. Motion of a the Moon around the Earth due to the deformation of space-time. Credit: Stanford University.

happens and a good way to summarize the theory is that of John Wheeler: *matter tells space how to bend, space tells matter how to move.*

Secondly, time is not represented. Finally, Fig. 3.8 gives the impression that the depression under the Earth, that is, the curvature of space, appears to be produced by the fact that it is subject to gravity and for this reason is pulled downwards. In reality, the curvature that is observed in Figure 3.8 is gravity itself and is not caused by it. Given the geometric nature of gravity, it is natural to hypothesize that gravity, that is, the space-time deformation, also influences the propagation of light. If we have a luminous object that emits a light ray directed towards an observer and if there is a mass interposed between the light-emitting body and the observer, the light will change its trajectory due to the space-time deformation. To test this prediction, Einstein proposed to verify the deflection of light rays coming from a group of distant stars, due to the mass of the Sun. The first attempt to verify the prediction was made by Erwin Freundlich, who organized an expedition for the eclipse of 21 August 2014 in Crimea. The expedition was unsuccessful because the members of the expedition were captured by Tsarist troops during the war. The event played in Einstein's favor. At that time, in fact, his theory predicted a deflection of 0.87 arc seconds,[5] equal to that calculated by von Soldner (in a paper of 1801 published in 1804) using the corpuscular theory of light and that of Newton's gravitation.

Einstein made a second estimate publicized in the famous conference of 25 November 1915. This estimate, 1.75 seconds of arc, was about double than the previous estimate. Obviously, the mishap that happened to Erwin Freundlich benefited Einstein a lot, since if the mission had been successful, the results would have contradicted his theory. Nonetheless, Einstein, who was not a "saint," in the following years showed in various ways his ingratitude towards Erwin Freundlich.

The second attempt to verify "the most beautiful of theories" was that of the director of the Cambridge Observatory, Arthur Eddington, who organized a double expedition to the island of Prince in Africa, and to Sobral, Brazil. In addition to the desire to verify the theory, the expedition had a more practical goal, that of avoiding Eddington's call under arms. As a pacifist and conscientious objector, he would certainly have refused to enlist and this would also have been an embarrassing situation for Trinity College, for the Cambridge Observatory and consequently for English science. The royal astronomer Frank Dyson pointed out to colleagues that on May 29, 1919, there would be an eclipse that would cross the Atlantic. The event was particularly interesting because Einstein's theory could be verified, since the Sun and the Moon would be in the Taurus constellation, in whose center the cluster of the Hyades is located. It was an almost unique opportunity and they could kill two birds with one stone: verify Einstein's theory, and solve the "Eddington" problem. Under Dyson's push, everything was organized and the two expeditions left, aboard the *Anselm*, Her Majesty's ship.

---

[5] More precisely, the value was 0.83 second of arc, because there was a mistake in the calculation.

Despite the weather problems, the two missions took photographs used by Eddington to determine the value of the deflection, after discarding data that strayed from his expectations. The angular deviation predicted by general relativity is tiny, corresponding to the angle subtended by a finger at a distance of 1 km. The result was declared in accordance with the predictions of Einstein's theory. Often the effect of light deflection is referred to as *gravitational lensing effect*, although more precisely it is the gravitational *deflection of light*. Although the deflection and the lens effect are based on the same principle, from a historical point of view they must be considered as two separate effects. By *lensing*, we usually indicate the distortion of distant galaxies, the formation of multiple images, or rings due to the presence of an object between the observer and the source. The first *lensing effect* study was carried out by Orest Chvolson in 1924 and by Einstein in 1936. In reality Einstein had already derived the effect in 1912, three years before the publication of the theory of general relativity, as a consequence of the deflection of light from gravitational fields, but the result had not been published.

Einstein brought with him to Berlin the results of the calculation, where he met the astronomer Freundlich. He discussed some possibilities with him to test his ideas. In 1936, he received in Princeton a visit from a Czech dishwasher, Rudi Mandl, who discussed the gravitational lens effect with Einsten. According to Mandl starlight focused by a gravitational lens could have influenced human evolution, producing important genetic mutations.

Under Mandl's pressing request to publish an article on the lens effect, Einstein, who initially refused because the effect in his opinion was negligible, published an article in *Science*. Curiously, in the article he explicitly wrote that he had published the article to satisfy Mandl's desire. Zwicky was the first to indicate that galaxy clusters could behave like gravitational lenses. This prediction was made in the 1937 article in which Zwicky had applied the virial theorem to the Coma cluster. More than 40 years had to pass before his intuition and the predictions of general relativity were verified in 1979, when an object image, doubled because of lensing, was observed: the twin quasar SBS 0957 + 561. Further confirmation of the gravitational lens effect comes from the observation of gravitational arcs produced by the distortion of the images of galaxies behind a cluster by Lynds and Petrosian.

In case of source, lens, and observer alignment, the following events can be observed:

- four copies of the source, called the Einstein cross effect. An example of the effect is produced by the galaxy ZW 2237 + 030 and the Quasar G2237 + 0305 placed directly behind it (Fig. 3.9)
- or a ring, the *Einstein ring*, the first discovered in 1998: B1938 + 666 (Fig. 3.10)

The observation of rings or other structures depends on the alignment between the distant object, the lens, and the observer (Fig. 3.11).

The arches, and the rings produced by the lensing effect are simply cosmic mirages, optical illusions. They are similar to the mirages observed in the deserts, in that case produced

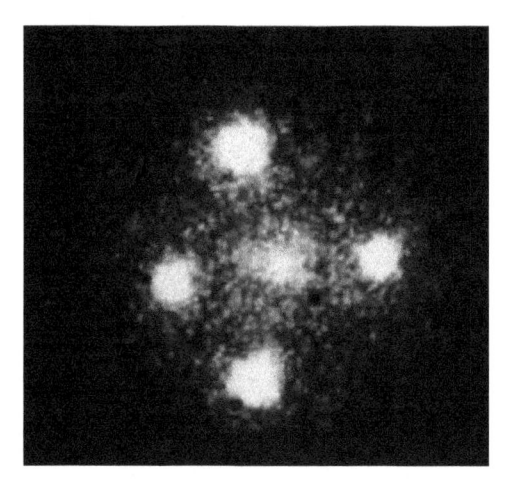

Fig. 3.9. Einstein cross: produced by the galaxy ZW 2237 + 030 and the Quasar G2237 + 0305. Credit: ESA/Hubble and NASA source.

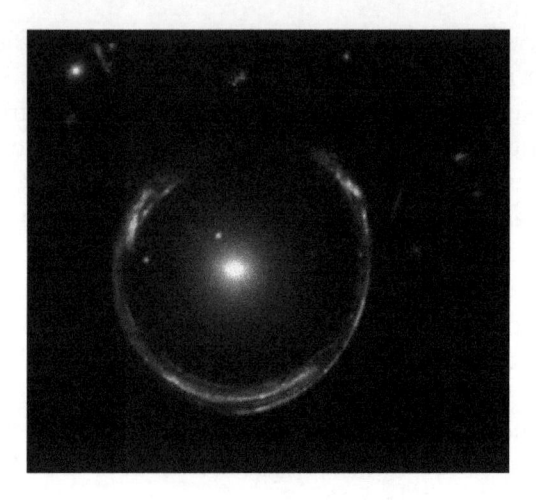

Fig. 3.10. Einstein's ring.
Source: ESA/Hubble NASA.

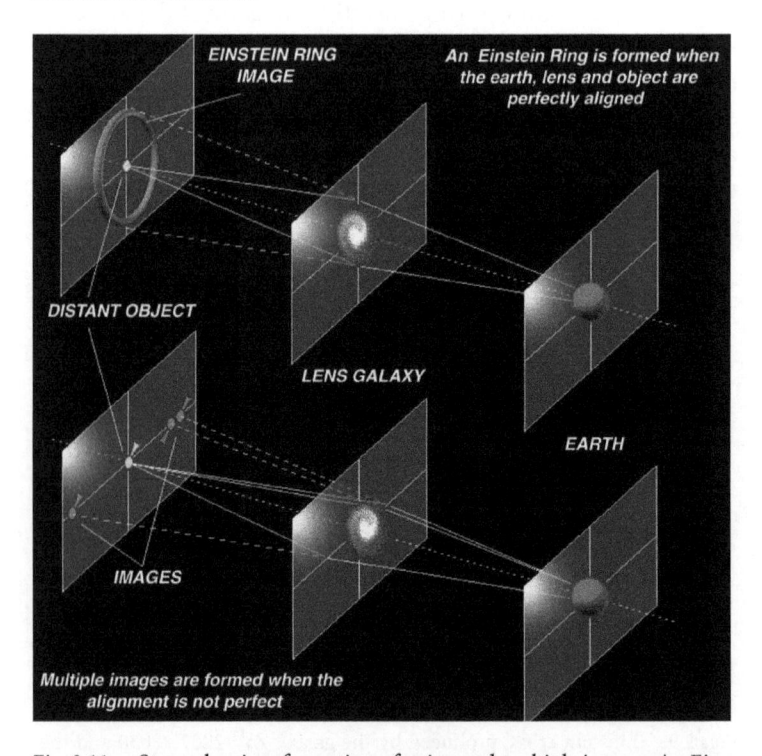

Fig. 3.11. Strong lensing: formation of a ring and multiple images. An Einstein ring is formed when the alignment between the Earth, the galaxy that creates the lens, and the distant galaxy is perfect. Credit: University of Manchester/Alastair Gunn.

by the temperature differences between the ground and the air, which deflect the light.

The lensing effect occurs naturally in three forms: *strong*, *weak*, and *microlensing effect*. The difference between them is the intensity of the effect. In the case of a strong lensing, arcs, rings, and the like can be observed, while in the weak one the effect is much less evident. The importance of the lensing effect is due to the fact that from the distortion of distant galaxies it is possible to determine the total mass of the lens, that is, sum of the stellar mass, gas, and dark matter. By measuring the stellar mass and that of the gas, the mass of dark matter can finally be determined.

The weak lensing effect was observed in 1990 by Tyson and collaborators, who identified a coherent alignment of the ellipticities of blue galaxies behind the Abell 1969 and CL 1409 + 52 clusters. Despite the weakness of the effect, using particular mathematical techniques and simulations it is possible to reconstruct the mass distribution of the lens. Figure 3.12 shows, on the left, multiple images of a blue galaxy behind the CL 0024 + 1654 cluster and on the right the false color computer reconstruction of the mass distribution of CL 0024 + 1654. The peaks represent the contribution of galaxies to the mass. The image shows how

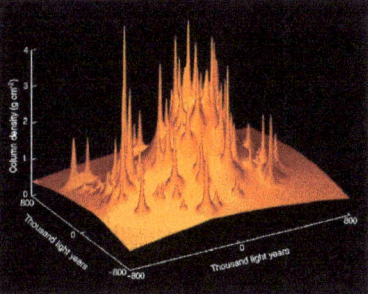

Fig. 3.12. Left: Gravitational arcs produced by the image of a galaxy behind the cluster CL 0024 + 1654. Right: Reconstruction of the mass of CL 0024 + 1654 in false colors. The plot represents dark matter per unit area. Pinnacles represent galaxies. Credit: Greg Kochanski, Ian Dell'Antonio, and Tony Tyson (Bell Labs).

The Invisible Universe

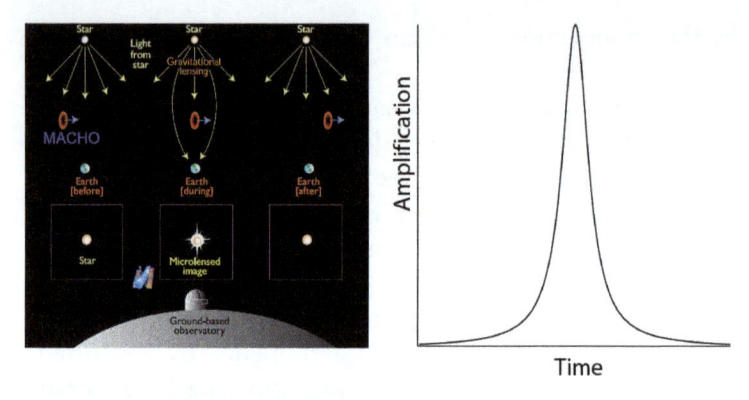

Fig. 3.13. Microlensing. Left panel: The MACHO moves until it cut the line between the observer and the star. The star light is amplified. Right: Light curve, amplification of the light from the star. Credit: Modified work by NASA, ESA, and A. Felid (STScI).

most of the mass is located between the galaxies, and is of the order of 40 times greater than the visible mass of the system. Lensing is now used routinely to reconstruct the distribution of dark matter and can be considered a method of "seeing" dark matter.

The last manifestation of lensing is *microlensing*, or the gravitational focusing produced by stellar or planetary objects. It had been proposed by Petrou in 1986 and by Paczynski in 1986 for the study of dark matter in the form of compact objects (see Chapter 5). In Fig. 3.13, is shown how the light of a star changes when a planetary object passes in its line of sight.

It is interesting to note that the Fermat principle, used to find the Snell–Descartes law, can be applied to the gravitational field (curvature of space-time) obtaining the deflection of light or more generally the so-called *gravitational lens equation*.

3.9 The Bullet Cluster

Clusters of galaxies constitute the largest gravitationally bound structures in the universe. Clusters are still being formed and are still growing thanks mainly to interactions and mergers with other galaxy clusters. They are made up of hundreds or thousands of galaxies, by diffuse hot gas that emits X-rays through which it is possible to trace the structure of these huge cosmic megacities. Clusters can collide with each other and by studying the result of the clash one can obtain clues about the properties of galaxies, and the nature of the mass that makes up the colliding clusters. The *bullet cluster*, 1E 0657-56, is located in the Carena constellation, and is the result of the collision between two galaxy clusters, a larger one with a mass of $2 \times 10^{15} M_\odot$ and one with a mass about thirty times less, at very high speed (5000 km/s), which occurred 100 million years ago. The distribution of the masses that make up the cluster of galaxies can be determined by studying the emission in the X band and the effect of the gravitational lens. Once the mass distribution has been reconstructed, we are faced with a unique spectacle: the baryonic mass emitting in the X-rays part of the electromagnetic spectrum is located in the central area of the cluster and is separated by two regions that contain most of the mass. In the collision, the gas, basically hydrogen, of the first cluster, collided with the gas of the second, forming shock fronts at the center of the cluster, imprinted in the X-ray emission of the cluster. Using lensing, one can reconstruct the distribution of the dominant mass in the cluster (blue area in Fig. 3.13), consisting of two mass concentrations located symmetrically with respect to the center. This mass has a completely different distribution from that of ordinary matter (red zone in Fig. 3.14).

The observations can be explained by remembering that dark matter has weak interaction with common matter. In the collision, the dark matter of each cluster continued its course undisturbed, forming the two concentrations far from the center, while the gas of the two clusters was slowed down by electromagnetic interactions and stopped in its central part. The cluster is therefore a very important indirect proof of the existence of dark matter, even if alone, 1E 0657-56 does not tell us if dark matter is made of particles or if it is made up of macroscopic and non-luminous objects such as, example, planets like our Jupiter. Another important thing is that while

Fig. 3.14. Composite image of the Bullet cluster (X-ray, Optical, and lensing map). The Bullet cluster is the collision between two galaxy clusters observed by Hubble. The blue areas are the dark matter of the two clusters, the red area the gas. Left panel: The cluster. Right panel: Dark matter and gas are indicated. Credit: X-ray: NASA/CXC/CfA/M. Markevitch. Optical and lensing map: NASA/STScI, Magellan/U. Arizona/D. Clowe. Lensing map: ESO WFI.

dark matter can give an explanation of the structure of this cluster, the theories of modified gravity, theories of gravity other than Einstein's, do not succeed.

We have seen that there is a series of indirect evidence of the existence of dark matter based on the study of galaxies, clusters, gravitational lenses. There are several other evidences of its existence, such as the way in which galaxies are distributed in space. In the next chapter we will see how the CMB gives further evidence of the existence of dark matter and provides a measure of how much dark matter there should be in the universe.

Chapter 4

THE HARMONY OF THE WORLD

Music is an embodiment of the harmony of the universe.

— Albert Jess

Another indirect evidence of the existence of dark matter is the CMB, the residual radio waves of the Big Bang. Many of us have observed this radiation on the screen of non-tuned old (non-digital) televisions. The television screen shows the so-called snow or white noise effect, and the CMB contributes to about 1% of that effect. CMB is one of the main tests of the Big Bang theory and tells us that the universe was extremely hot in the past and that it then cooled and expanded. Buried in the details of this radiation there are other indirect evidences of the existence of dark matter, as we will see shortly.

4.1 The Music of the Universe

On the CMB map, we can observe areas with different densities (density fluctuations). These fluctuations are basically sound waves. Sound waves do not propagate in empty space, so how could sound waves exist in the universe at recombination? The answer is that at the time of recombination the space was not as empty as it is today, but was full of dense plasma. This explains how sound waves could be generated and propagate. It is therefore not strange that the primordial universe contained sound waves, because if we think about it, the sound waves that we experience in everyday life are nothing but perturbations, or fluctuations, of pressure and density that propagate in the air. In the universe, the propagation medium was plasma.

In the primordial universe there were quantum fluctuations in the field that produced inflation. The latter, by rapidly producing the expansion of the universe, transformed quantum fluctuations into fluctuations of approximately equal amplitudes on all scales. Then inflation filled up universe with sound waves. In addition, the density fluctuations produced by inflation were almost all created at the same time. More precisely, the harmonic structure regularity implies that the oscillations of the density fluctuations having a given size reach their maximum compression and rarefactions at the same time. Then, their oscillations started simultaneously. This produced a *sound spectrum* (see Fig. 4.5, left) similar to that seen in common musical instruments (Fig. 4.5, right). To better understand, let's study how sound is generated in a flute like the one shown in Fig. 4.1.

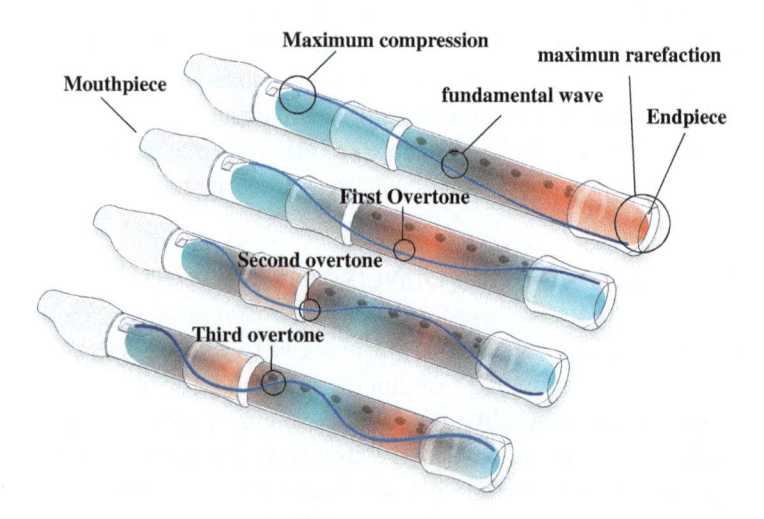

Fig. 4.1. Harmonics in a flute. The fundamental wave has a maximum compression in the mouthpiece (blue zone) and maximum rarefaction in the final part of the instrument (red zone). In addition to *the fundamental wave*, there are waves with half length (first harmonic), a third (second harmonic), a quarter of the fundamental wave. The so-called *harmonics*. Credit: Modification of work by Scientific American 290N2 44 (2004), and Bryan Christie.

When we blow into a flute, we make the air contained in it move and swing, and waves are generated. The so-called *fundamental frequency* (Fig. 4.1) corresponds to a wave with maximum air oscillations at the two extremes (mouthpiece and final part) and minimum in the central part. Near the mouthpiece there is the maximum compression of the air, and in the final part the maximum rarefaction. The length of the tube that constitutes the flute is equal to half a wavelength. That is, the distance between a zone of maximum density and one of minimum density is equal to half a wavelength (Fig. 4.2).

In addition to the *fundamental frequency*, other "secondary frequencies" are generated which are called *harmonics*. The wavelengths of the latter are whole fractions of the fundamental frequency. The first harmonic has a wavelength that is half the fundamental frequency. The second harmonic has a wavelength one third of the fundamental one, and so on. While the fundamental frequency determines the perceived pitch of the sound, the harmonics determine its *timbre*, that is, the quality of the sound. The waves present in the universe are similar to those produced in an instrument, with the difference that in the universe the waves are oscillations over time.

Continuing with our analogy between waves of an instrument and those in the universe, we can say that the length of the tube represents the time it takes for the wave to reach recombination, starting from the inflation. A fundamental thing to note is that while in the case of a musical instrument the periods of oscillations of the waves are in the range between

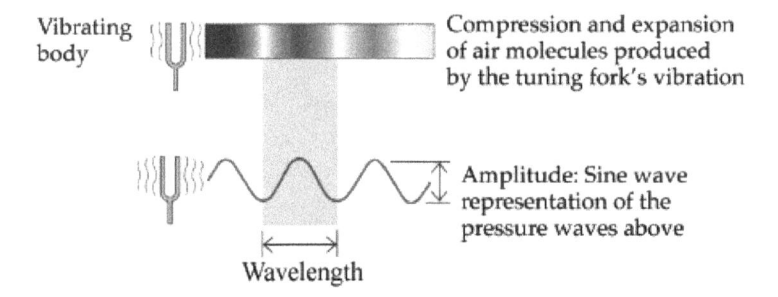

Fig. 4.2. Areas of maximum density spaced by a wavelength.

millionths and thousandths of a second, the fundamental tone in the universe had made half an oscillation in 380,000 years, the time that elapses between inflation and recombination (380,000 years later).

Now consider Fig. 4.3 and compare it with Fig. 4.1. In Fig. 4.3, each strip represents the change in temperature and density due to the propagation of harmonics in the plasma. The first strip, at the top, represents the fundamental frequency. At the time of inflation, corresponding to the mouthpiece of the flute, we have the maximum density (and temperature). At recombination, corresponding to the final part of the flute, we have the minimum temperature or density (maximum rarefaction in the flute). As in the flute, harmonics have lengths that are whole fractions of the fundamental. Therefore, passing to the second, third, and fourth peak (harmonics), at recombination, the regions with maximum or minimum temperature are always smaller, and there are different maximum and minimum temperatures.

If in the universe we had only one musical note and we built a map similar to that of the background radiation, we would see that it is made up of denser areas and less dense areas separated by half a wavelength. The map would not be like that

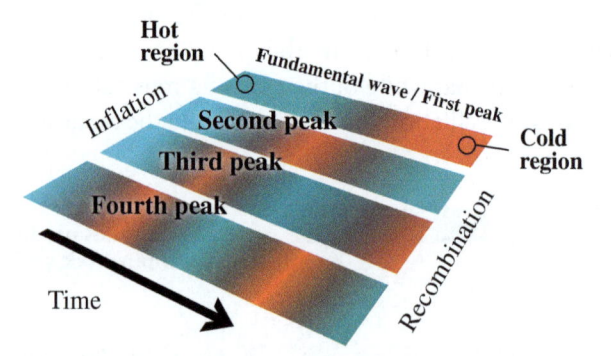

Fig. 4.3. Oscillations of sound waves in a plasma. The waves produce compression and rarefaction of different regions. The fundamental oscillation produces a maximum of temperature (blue zone) for inflation and a minimum for recombination (red zone). The harmonics produce different regions with maximum and minimum temperatures. Credit: Modification of work by Scientific American 290N2 44 (2004), and Bryan Christie.

of the background radiation seen in the left panel of Fig. 4.4, but regular as a checkerboard (Fig. 4.4, right panel), with two boxes, corresponding to the maximum and minimum density, separated by half a wavelength.

However, even the note of a musical instrument is not only made up of a dominant (fundamental) tone, but multiples of it, the harmonics. Consequently, the background radiation map is more complex than the checkerboard map because it is generated by more than one sound, or in the presence of harmonics, as seen in the left panel of Fig. 4.4. An analysis of the background radiation map by determining the sound spectrum, that is, the content of each wavelength in the map, reveals that half the wavelength of the fundamental tone occupies an angle of 1 degree in the sky (Fig. 4.5, left panel). This is the average distance between the maximum density

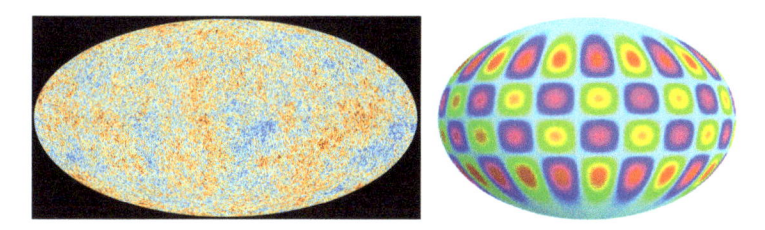

Fig. 4.4. Left: Map of the PLANCK CMB. Credit: ESA and the Planck Collaboration. Right: Map generated by a single wave.

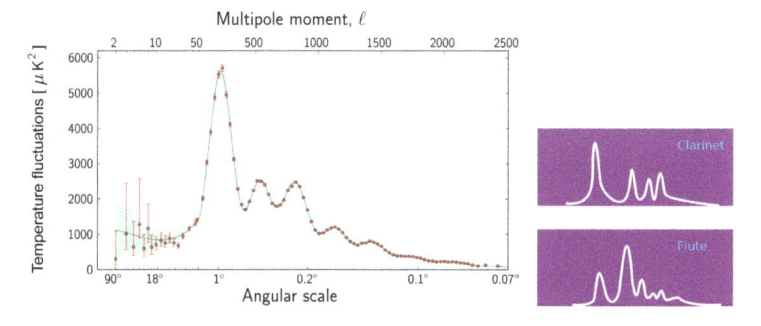

Fig. 4.5. Left: Universe sound spectrum, from PLANCK satellite. Credit: ESA and the Planck Collaboration. Right: Sound spectrum of some instruments. Credit: Modification of work by Judy Brown (https://acoustics.org/pressroom/httpdocs/139th/brown.htm).

(blue) and the minimum density (orange) points. The plot in the left panel of Fig. 4.5 represents the sound spectrum of the universe, and it is incredibly similar to that of musical instruments (Fig. 4.5, right panel). In both cases there are a series of acoustic peaks. If you look at the spectrum of the clarinet, you can see that it is almost identical to that of the primordial universe: a flat area, a climb to a peak, and a series of successive peaks with lower heights. The system of peaks of the sound spectrum of the background radiation was hypothesized in the '60s of the last century by Sakharov, Sunyaev and Zeldovich, and Peebles and Yu, long before the peaks were observed.

In addition to the fundamental tone, there are several harmonics with shorter wavelengths and smaller amplitudes. If the distance of the *surface of last scattering* is known, we find that 2 degrees in the sky, the wavelength of the fundamental tone, correspond to 1 million light-years. The universe was, therefore, crossed by sound waves of enormous wavelength.

In other words, the universe was a sort of huge cosmic musical instrument. In the universe there were sound waves but the oscillations occurred in very long times, and therefore, even if we had been present, the sounds would not be perceptible to human ears. However, there is a way to make the sounds of the primordial universe audible, processing the data so that they become perceptible to our ears. In this way it is possible to obtain the sound of the primordial universe. This is what John Cramer of the University of Washington (Seattle) did, and if you want you can hear it at the link in the footnote.[1]

As noticed by Alberto Casas,[2] the sound produced by a musical instrument can be used to get information about the instrument, such as what material it is made of (Fig. 4.5, right panel). In the same way, the sound spectrum of the primordial universe allows us to determine its matter content, its geometric shape and so on.

---

[1] https://www.youtube.com/watch?v=LtpXbrpEU3c
[2] "Materia oscura. L'elemento più misterioso dell'Universo" ("Dark Matter. The most mysterious element in Universe"), RBA editor.

How is this possible?

The sound wave of the fundamental wave produced after half an oscillation an increase in density of the plasma, which therefore underwent a compression. This compression was accentuated by the presence of the force of gravity, which in turn depends on the quantity of matter present: more matter, more gravity.

The first peak is generated by the compression of a region that reached its maximum compression at the time of recombination. The second peak (Fig. 4.5, left panel) corresponds to the sound wave that completed an oscillation at recombination. During the oscillation, the matter first compressed and then expanded under the action of the force linked to the pressure of the plasma, reaching its maximum rarefaction at the time of recombination. The final result was a lower peak with respect the first one.

The third peak is generated by an oscillation that reached its second maximum compression at the recombination epoch.

There is another reason why peaks have decreasing amplitudes. The recombination process is not instantaneous. The scales associated with the distances that photons can travel in that time interval are "diluted." In other words, photons in their motion eliminate density fluctuations on a scale of 10 Mpc. This scale is called the *Silk scale*, and the effect is named *Silk damping*, from the name of the British astrophysicist who described the phenomenon in 1968. The mass associated with this scale is the minimum mass that an object must have in order to collapse after recombination, and is equal to the typical mass of a galaxy. The decrease in the height of the peaks, expecially the last ones, is produced by the Silk damping.

The position of the peaks in Fig. 4.5 (left panel) is sensitive to the universe curvature, while their shape, and height, is sensitive to the density of matter. The position of the first peak is consistent with a flat universe. The height of the second peak relative to the first is related to baryons content, while that of the third peak relative to the first peak is related to the density of dark matter. The height of the second peak gives us the percentage of ordinary matter, which is about 5%, and therefore

26% of the universe is made up of dark matter, the difference between total matter (31%) and baryonic matter (5%).

4.2 Sound Harmonics and the Geometry of the Universe

The peaks in the background radiation sound spectrum (Fig. 4.5, left panel) can also be used to determine what kind of space we live in. We have seen that the average distance between orange and blue areas in the map of the background radiation (Fig. 4.4, left panel) is equal to a half wavelength of the fundamental tone, corresponding to an angle of about 1 degree, as confirmed by the position of the first peak in the sound spectrum. In the theory of general relativity, space-time is curved by the presence of matter and energy. Depending on the material content, the universe, as described in Chapter 1, can have three different geometries: closed universe if $\Omega > 1$, flat universe if $\Omega = 1$, hyperbolic universe if $\Omega < 1$.

As shown in Fig. 4.6 (left panel), if we were in a universe with positive curvature, the visual angle that would occupy half the wavelength of the fundamental tone would be greater than 1 degree, and in a universe with negative curvature it would be less than 1 degree (Fig. 4.6, right panel). From this we deduce that our universe has zero curvature and is thus flat, that is, $\Omega_{Total} = 1$ (Fig. 4.6, central part). Recall, from Chapter 1, that an open universe has $\Omega_{Total} < 1$, and a closed universe $\Omega_{Total} > 1$.

Another way of determining the geometry of the universe is to simulate the characteristics of the CMB map in different universes and compare them with the observations of the same (see Chapter 9.3). Like the method discussed above, the comparison confirms that the universe is flat and that the sum of ordinary matter (4.86%) with photons (0.005%), neutrinos (0.4%), and dark matter (25.89%) constitutes approximately 31% of the critical density. A large percentage of mass is missing to reach the critical density, namely 69%. This matter can be neither ordinary matter nor dark matter, and it is called *dark energy*.

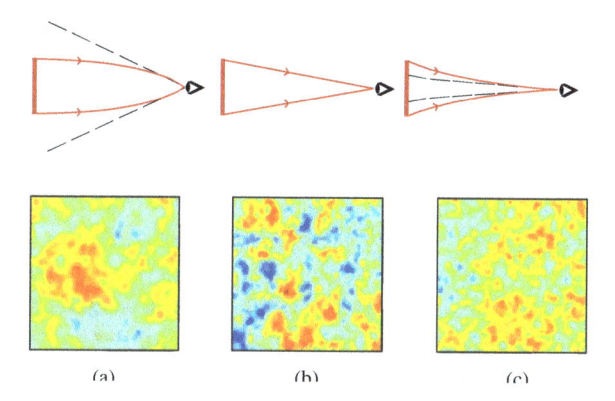

Fig. 4.6. Inhomogeneity in the CMB and geometry of space. In the top three figures, the thick brown lines represent the fundamental wavelength, the thin brown lines the way light propagates in closed, flat, and open universes, and the dashed black line the angle subtended by the geometry we consider. In a flat universe the fundamental wavelength subtends an angle of 1 degree. In a closed universe the angle subtended is larger (panel a) (>1°), while in an open universe (panel c), the angle is smaller (<1°), with respect to a flat universe (1°) (panel b). Then cold and hot spots appear larger in a closed universe, and smaller in a open universe, with respect to a flat universe. This is visible in the bottom part of the figure. Credit: Cosmology Group/Lawrence Berkley Laboratories.

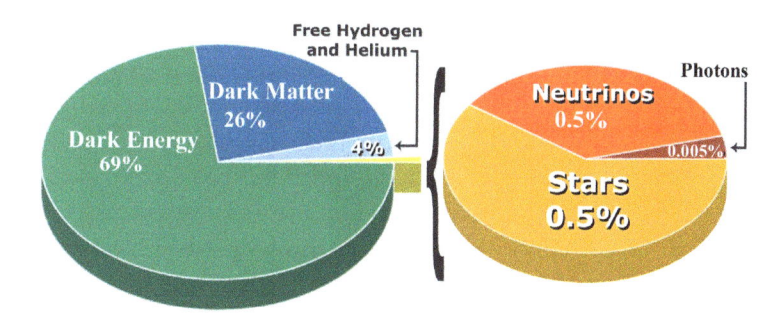

Fig. 4.7. Cosmic inventory: content of dark matter, dark energy, and other components in the universe.

The peaks of the CMB shown in Fig. 4.3 (left panel), and the whole curve, can be reproduced using a model with 6 numbers or parameters called *cosmic concordance model*, or ΛCDM *model*, or the *standard cosmology model*. In this model, our universe has a flat geometry, it contains ordinary matter, dark matter, and dark energy in the said percentages. In the model, the "Λ" refers to dark energy, and "CDM" to dark matter (cold dark matter).

4.3 Cosmic Symphonies and Dark Matter

As we will see in Chapter 6, the CMB gives us information on the nature of dark matter. For example, we know that it must be cold, that is, the dispersion velocity (random speed) of the particles must be much lower than the speed of light, since the *warm dark matter* made by neutrinos generates perturbations different than those necessary to form cosmic structures.

Another important information that comes from the homogeneity and isotropy of the CMB is that on scales larger than 100 Mpc the universe is homogeneous, that is, the average density of matter is everywhere the same, and isotropic, that is, equal regardless of the direction in which one looks.[3]

Background radiation is also a test of inflation theory. The density fluctuations that are observed on the CMB were generated by quantum fluctuations produced during inflation, amplified by gravity to become macroscopic. The first detailed calculations on primordial fluctuations were made in early 1981 by Mukhanov and Chibisov and by Hawking, Starobinsky, Turner, and Guth in 1982. Inflation provides a precise spectrum for fluctuations, which has been confirmed by the CMB study.[4] Another of the inflation forecasts is the production of primordial gravitational waves, which have left a mark on the CMB map, called *B polarization modes*. The amplitude of these disturbances could be a direct measure of the energy scale at which inflation occurred, thus indicating the energy of the particle, the *inflaton*, which originated it. They can be also used to distinguish between the Big Bang model and the cyclic models described in section 10.5.

---

[3] To simplify the calculations, the hypothesis of homogeneity and isotropy, dubbed the *cosmological principle*, was used by Einstein and other physicists.
[4] Technically we speak of *spectral index, Gaussianity, adiabaticity*.

In 2014, in an experiment in Antarctica, carried out with the BICEP2 instrument, the signals of the generation of primordial gravitational waves were apparently observed, but later the claim was withdrawn since detection was disproved by the studies from the PLANCK satellite.

Although most of the features of CMB have been understood, others are not. Data obtained with WMAP and PLANCK have shown that the CMB anisotropy maps reveal different anomalies compared to statistical isotropy, in contradiction with the cosmological principle. One of these anomalies is the presence of a cold spot of angular size about $5-10^0$ (Fig. 4.8). This area could be due to a large void located between us and the cold spot direction or to other more exotic reasons. According to Laura Mersini-Houghton, the cold spot could be due to a *quantum entanglement*, a primordial correlation between our universe and another, later separated by inflation expansion.

In conclusion, the CMB is a powerful laboratory for cosmology. Great observational efforts have been made to try to reveal the signals of primordial gravitational waves. The answer could come from experiments such as BICEP and QUIJOTE or from future satellites that have been proposed: PIXIE, LiteBIRD, JAXA, CORE. A deeper study of the CMB will provide us with new ways to study the thermal history of the universe, inflation, and dark matter.

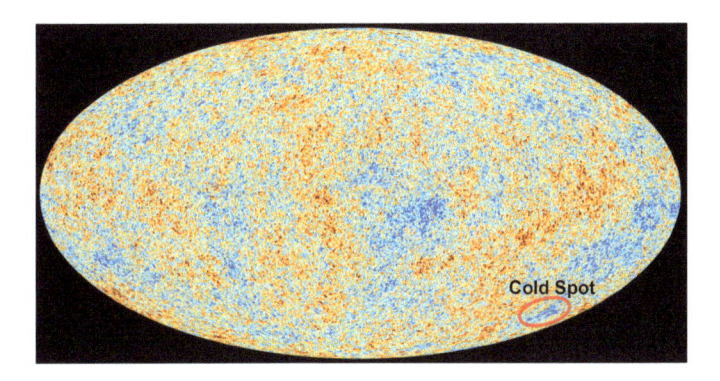

Fig. 4.8. The cold spot in the CMB.

Chapter 5

WHAT IS DARK MATTER?

The universe was built according to a project whose profound symmetry is somehow present in the internal structure of our intellect.

— Paul Valery

5.1 Ordinary Dark Matter?

We have discussed a series of indirect tests, all converging on the claim that our universe must contain dark matter. The open question remaining is that of understanding what dark matter is made of. The first idea that can come to mind is that it is usual matter that for some reason is not visible, for example because the radiation emitted is so weak that it is not detectable. In 1956, Heeschen searched for neutral hydrogen (HI) emissions in the Coma cluster, finding it, but three years later Muller showed that Heeschen's result was not correct.

In the mid-1960s, H. Rood tried to understand what the dark matter within galaxy clusters was made of. He came to the conclusion that this matter should be found in the intergalactic space. Since neutral hydrogen had not been observed, people began to think that it could be ionized hydrogen. Observations in the radio and X bands made it possible to obtain estimates of how much gas the clusters contained, coming to the conclusion that it should be only 2% of the material that kept the cluster gravitationally bound. Too low a quantity to identify gas with dark matter. Once these possibilities were discarded, attention was paid to

- collapsed massive objects having masses from hundreds of millions to thousands of billions that of the Sun — hypothesis rejected because of the absence of tidal distortion of galaxies
- neutral hydrogen snowballs — possibility never discarded
- dwarf stars, planets, brown, red, white dwarfs, or black holes

Kim Griest introduced the acronym MACHO (massive astrophysical compact halo objects) for these objects.

The idea that dark matter was made by MACHOs seemed like an excellent intuition, and for this reason, experiments were designed to test the hypothesis. We will now discuss the *MACHO dark matter* constituted by objects related to stellar formation and evolution. We will see them one by one.

5.2 The Stars

The Sun is a star eight light-minutes away from us. It is the source of the energy necessary for life on Earth. The Sun is a medium to small star with a radius more than a hundred times that of the Earth, a mass 300,000 times greater, and density similar to that of water. The Sun, like the other stars, is basically a plasma ball whose stable structure is due to the balance between the gravitational force, which tends to make it collapse, and the gas pressure, generated by the central nuclear reactions which transform hydrogen to helium when the star is young. Nuclear reactions produce electromagnetic radiation, a flow of elementary particles called *solar wind* and *neutrinos*. In the case of the Sun, the surface have a temperature of 5780 K, and it grows, moving towards the center, up to tens of millions of degrees. Aside from the central area of the star, where nuclear reactions are produced, going outwards the star is characterized by a *radiative and a convective region*. In the first region, energy is transported through radiation, and in the second it is transported by convective motions. The zones are positioned differently for small and large stars. The stars also have an atmosphere. Starting from the *photosphere* (from the Greek "sphere of light") from which the rays that we see come up, we meet the *chromosphere* (literally, "sphere of color"). Then follows the *corona*, consisting of a rarefied gas with temperatures of a few million degrees.

The stars also have different colors depending on their surface temperature. The coldest ones, around 2000 K, are dark red, whereas the warmest ones (with temperatures equal or larger than 20,000 K), tend to be blue. Understanding the color differences is simple. In daily experience, when a blacksmith heats the iron, it begins to emit dark red light, and with time and the increase in its temperature, the emission first turns to yellow, and then to blue.

The internal composition of the stars changes with age. Stars like our Sun are made up of 70% hydrogen, 28% helium, and 2% heavier elements, and over time the amount of hydrogen decreases and that of the other components grows. There are stars of various sizes and masses. The smallest are those with the minimum mass to trigger the hydrogen fusion (0.08 solar masses), and the largest those reaching masses of the order of 100 solar masses. The dimensions can vary from 10 km, in the case of *neutron stars*, to billions of kilometers in the case of *supergiants*. Stars are formed by the gravitational collapse of large gas clouds and are formed into groups generating associations or star clusters. As we will discuss in Appendix B, in order to form a star or a galaxy, it is necessary to start from a gas cloud with mass larger than a critical mass dubbed *Jeans mass*, which takes its name from the British physicist James Jeans, who in 1902 published a study on the problem of the gravitational collapse of a gas cloud. Jeans showed that there is a limit mass, above which the cloud can collapse, and which is usually of the order of a thousand solar masses. As a consequence, stars are not formed individually but in groups of thousands. Unlike our Sun, stars usually have a partner. In addition to groups, stars are part of very large structures, the galaxies, made up of a few hundred billion stars, as in the case of our galaxy.

Even if the stars are formed by the collapse of large masses of gas and are formed in groups, each of them follows their own destiny. *Kant–Laplace's nebular hypothesis* is a theory that within its limits manages to give an idea of the formation of a star and the planetary system linked to it. It assumes that a mass of gas enters instability, for example due to the explosion of a nearby star, and collapses. During the collapse, the nebula begins to rotate faster and faster, due to the conservation of the angular momentum, and heats up. When the temperature reaches the value necessary for nuclear fusion, a star is formed in the center of the collapsing nebula. The force, which produces the rotation and pressure of the gas, gives rise to a flattened structure, the *protoplanetary disk*, with the *protostar* at its center. Planets are formed from the protoplanetary disk. During the formation of the solar system, at distances lower than that of Jupiter, the temperature was so high that it did not allow the condensation of molecules such as water or methane. In this region, planets formed by accumulation of rocky objects, called *planetesimals*, which colliding with each other gave rise to terrestrial planets. Beyond the *snow line*, at a distance of about 3 astronomical units, the temperature was low enough to cause the planetesimals to capture gas, giving rise to planets such as Jupiter (318 times more massive than the Earth) and Saturn. Uranus and Neptune captured less gas and condensed around ice cores. At the end of this process, the solar wind blew away gas and dust, blocking the planetary growth process. Until 1995, no other solar system, except our own, was known. In that year Mayor and Queloz, two Swiss astronomers, indirectly observed a planet of the star *51 Pegasi*. For that, they received the 2020 Nobel Prize. Since then, with the improvement of observation techniques and thanks to the help of space missions such as *Kepler*, the number of *extrasolar planets* has increased to several thousand and it is estimated that there should be a billion in our galaxy. The first planets discovered had masses greater than those of Jupiter, because the observational technique favored the discovery of massive planets. Today, planets similar to our Earth have been discovered. The most similar to Earth having

ESI (Earth Similitude Index) 0.90 is *K2-72*, 228 light-years away from Earth, with mass 2.7, and radius 0.82 times that of the Earth. Other Earth-like planets are those revolving around the *Trappist* star, and in particular the planet *Trappist-1*. This planet is located 39.5 light-years from Earth, and it has a mass of 0.62, and a radius of 0.92 that of Earth. Among the other planets, *Kepler-442b* is 111 light-years away. In addition to the similarity with the Earth, it is probably located more in the center of the *habitable zone*[1] of its star.

5.3 Death of Stars and Cosmic Fossils

We often think of the transience of human life by comparing it with the life of stars, galaxies, the universe. In reality, although stars, galaxies, and the universe have an immeasurably greater life than ours, they too are destined to die. As for the universe, the possible ends are described in the last chapter.

Like men, stars live and are destined to die, and like men, they die in different ways. The parameter that determines the type of death for a star is its mass. Stars with masses similar to that of the Sun consume their fuel (hydrogen) slowly. They can live up to 10 billion years. When hydrogen has been transformed into helium, the star will contract, triggering hydrogen fusion reactions in spherical shells outside the nucleus. When these reactions also end, the star will re-contract and trigger helium fusion, with a central temperature of 100 million degrees. The Sun will expand, in the phase of a *red giant*, until it touches the Earth. When even the helium will no longer act as fuel, the Sun will contract, but this time the contraction will not be sufficient to trigger the carbon fusion. The collapse will expel the external parts that will form a planetary nebula, and at its center a *white dwarf* will be formed. The force that stops the gravitational collapse, forming white dwarfs, the pressure of the *degenerate electrons*, a particular phase of matter characterized by high density and

---

[1] The habitable zone, also named *Goldilocks Zone*, is the region in which a planet has liquid water on its surface.

high kinetic energy, which originates from the *Pauli exclusion principle*. This principle was introduced in 1925 by Wolfgang Pauli. It says that two fermions, for example electrons, of the same species cannot occupy the same state, that is, they cannot have the same energy, angular momentum, and momentum. When particles get closer, they tend to occupy states with larger energy.

The Pauli principle forces electrons to move frantically and occupy states with high kinetic energy. These motions generate a pressure that manages to balance the force of gravity, allowing for the existence of the white dwarf. The high density and velocity state of the electrons is called a *degenerate state* and the electrons are called *degenerate electrons*.

Stars with masses greater than 8 solar masses burn their fuel faster and live 10 million years. In a series of successive contractions and expansions, they will be able to trigger all the elements less massive than iron. Iron fusion requires more energy than it emits. Consequently, nuclear reactions will stop. The star will explode giving rise to a *supernova*, which will eject the shells of external material into space, while at the same time the core will be compressed to form a *neutron star*, when electrons and protons are fused. Stars with larger masses give rise to *black holes*.

5.4 Brown Dwarfs

Brown dwarfs are celestial objects with masses intermediate between 1, and 75–80 Jupiter mass. This is the minimum mass necessary to fuse hydrogen. Objects with masses of about 13 Jovian masses are called *brown sub-dwarfs*. For masses in the range 13–65 Jovian masses, the reactions of deuterium inside the brown dwarf are triggered, and above 65 Jovian masses, those of lithium. These stars have a temperature of about 2000 K, much lower than that of the Sun which is 5780 K. Given their small mass, brown dwarfs are unable to trigger hydrogen fusion reactions, but are able to generate energy thanks to the fusion reactions of deuterium and lithium. These elements are easier to fuse than hydrogen, and for this reason they are absent in normal stars that burn them in a short time. In addition to these reactions, brown dwarfs have another source of energy: gravitational contraction. Not far from us, at 6.5 light-years, there is a binary system of brown dwarfs: *WISE 1049-5319*.

5.5 Cosmic Fossils

Other objects that could constitute baryonic dark matter are remnants of stellar evolution, namely white dwarfs, neutron stars, and black holes.

A white dwarf is a low-luminosity star with dimensions similar to that of Earth but masses similar to those of the Sun. Consequently, these objects are very compact, with a gravitational field a million times greater than the terrestrial one and a density a million times greater than that of water. They are the final phase of the evolution of stars of mass similar to that of the Sun and, after the *main sequence phase*, follow an evolution that leads them to instability phases. The outer mass shells are expelled and give life to a *planetary nebula*, leaving a white dwarf in the center. White dwarfs are formed when the gravitational collapse of the parent star is stopped by the pressure of the degenerated electrons.

Non-rotating white dwarfs can have a maximum mass of 1.44 solar masses, called the *Chandrasekhar limit*. When these stars originate, they have surface temperatures of the order of 100,000 K, which decrease over time until the star no longer emits radiation and is called a *black dwarf*. The time required for cooling is much greater than the age of the universe and therefore it is thought that there are still no black dwarfs in our universe. In our neighborhood (21 light-years), there are 8 white dwarfs of the 11,000 estimated in the universe.

The second possibility is given by neutron stars. Like white dwarfs, neutron stars are degenerate stars and have a size of about 10 km, masses between 1.4 and 3 $M_\odot$, although the largest observed have mass 2.01 $M_\odot$.

The density of these objects is similar to that of atomic nuclei, around 10^{14} g/cm^3, or 198 million tons per cubic centimeter, and a gravitational field one hundred billion (10^{11}) times greater than that of the Earth. The faster ones perform a complete rotation in periods of a thousandth of a second, and the slower ones in tens of seconds. Two years after the discovery of the neutron by Chadwick, Baade and Zwicky proposed the existence of stars entirely composed of neutrons. Neutron stars were discovered in 1967 by observations made in Cambridge and were observed for the first time by Jocelyn Burnell Bell.[2] These stars, called *pulsars*, emit radiation while they rotate. They can be observed only if the light is emitted in our line of sight, as extremely regular impulses. Pulsars are a kind of cosmic lighthouses.

Finally, if the mass of the progenitor stars is some tens of solar masses, there is no force in nature that can stop the gravitational collapse. The collapse of the star gives rise to a *black hole*. These objects are perfect laboratories in which to study the consequences of general relativity and quantum mechanics. They are characterized by a region, called the *event horizon* (Fig. 5.1), from which no object — not even light — can

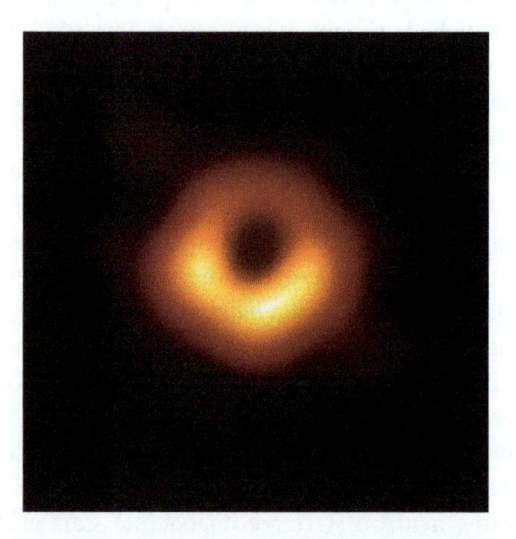

Fig. 5.1. Image of the black hole (event horizon) of M87 obtained with the data of the Event Horizon Telescope. Credit: Event Horizon Telescope team.

---

[2] It is interesting to note that Jocelyn Burnell Bell did not receive the Nobel Prize for her discovery, which was assigned to her supervisor Antony Hewish.

escape. If the Sun could collapse to form a black hole, its radius would be compressed to 3 km.

The black holes formed by the gravitational collapse of a star have masses from 5 to 10 solar masses, and exceptionally up to 20 solar masses. Black holes with mass around 30 solar masses were discovered in the collision that generated the first gravitational wave, observed in 2015 by the LIGO observatory (Fig. 5.2). In addition to these black holes, there are others of millions of solar masses in the center of galaxies like ours, or billions of solar masses in the center of elliptical galaxies like M87.

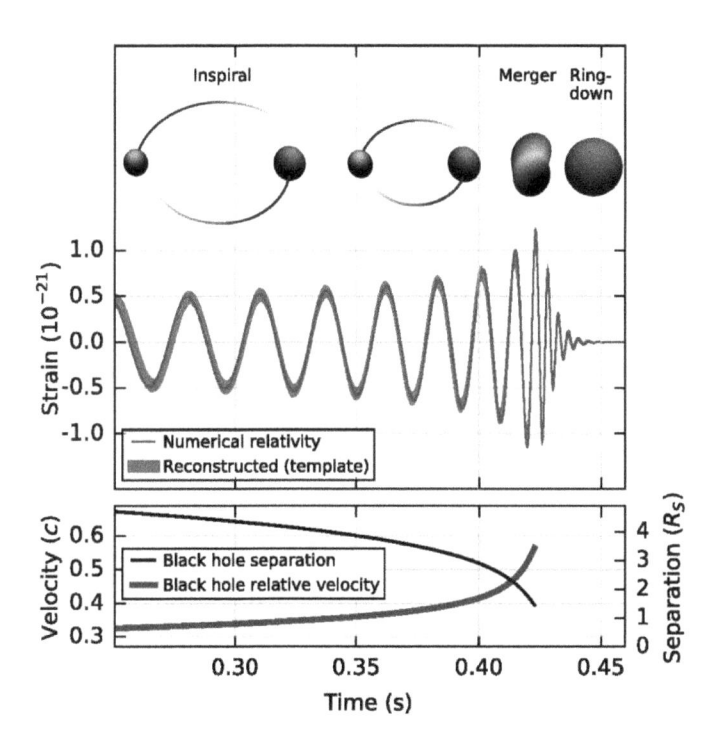

Fig. 5.2. First observation of gravitational waves in September 2015. Top panel: The event called GW150914, being a gravitational wave (GW), received on September 14, 2015, due to the merger of a 36 M_o black hole and a 29 M_o black hole. The top panel shows how the black holes approach each other, and fuse in one black hole. The oscillating line the gravitational-wave strain amplitude. Bottom panel: Black hole relative velocity, and separation. Credit: LIGO Collaboration (https://doi.org/10.1103/PhysRevLett.116.061102).

5.6 The Search for MACHOs

Although it is clear today that MACHOs cannot explain the totality of dark matter, two decades ago the situation was much less clear.

Nevertheless, many proofs against the existence of these objects — such as for example the predictions of the primordial nucleosynthesis, which foresees that there is 5% of baryonic matter in the universe — several studies and experiments were designed to establish what the quantity of these objects was. It was also clear in this case that the phenomenon of gravitational lensing could play an important role. In 1979 Chang and Refsdal showed, contrary to what Einstein had thought, that even stars could act as gravitational lenses. In 1986 Paczyński proposed this effect to search for compact objects in our galaxy, and Nemiroff made calculations on the probability of microlensing events due to these objects.

The microlensing effect is characterized by the increase in stellar light depending on the mass of the object, as shown in Fig 3.13. For example, if we look at a star in the *Magellanic Clouds* and a MACHO passes through the line that connects the observer with the star observed, a growth in stellar brightness will be observed (see Fig. 3.13). The times of variation range from a few hours to a year for MACHOs with masses between a ten millionth and a hundred solar masses. Three collaborations, EROS (Experience pour la Recherche d'Objets Sombres [Experiment for the search for dark objects]), OGLE (Optical Gravitational Lensing Experiment), and MACHO, began a search for MACHOs in the '90s. The MACHO collaboration, after about 6 years and after observing the light curves of tens of millions of stars, revealed 14–17 microlensing events. The conclusion was that the MACHOs should make up between 8% and 50% of the mass of the halo of our galaxy. After a year, the results led to a reduction in the percentage. The MACHOs could constitute at most 8% of the mass of the halo.

5.7 Primordial Black Holes

One last possibility of objects that could constitute baryonic dark matter is primordial black holes. There may exist black holes not generated by the collapse of stars, but originated in the first moments of the formation of the universe due to the very high material density during those phases. Primordial black holes have fairly small masses, $\sim 10^{12}$ kg, similar to those of a comet. In the first moments after the Big Bang, simple fluctuations in material density can result in regions so dense that they collapse to form black holes. The idea that primordial black holes could constitute all or part of dark matter has become an active study area after the detection, from the LIGO/VIRGO collaboration, of gravitational waves due to coalescence of black holes. The black holes that produced the 2015 gravitational wave had unexpected masses for a black hole. One possibility is that they constitute a large family of black holes with a spatial distribution similar to that of dark matter, originated by growth from the population of primordial black holes. The limits on baryonic matter given by primordial nucleosynthesis and CMB would not apply to such objects that formed before nucleosynthesis, at least for small mass objects.

The limits on primordial black holes obtained with the EROS collaboration and CMB anisotropies have been revisited, and some studies have shown that primordial black holes could constitute dark matter, while others have found opposite results. The matter is not very clear.

In conclusion, excluding primordial black holes, the theory of primordial nucleosynthesis puts limits on the mass of baryonic matter in the universe, which is 5%, as confirmed by the CMB study. Furthermore, only about 1% condensed to form stars, planets, and other compact objects. Most of the baryonic matter is found in interstellar matter, in the hot gas inside galaxies and clusters. Consequently, these limits indicate that dark matter, which constitutes 26% of the matter of the universe, cannot be made entirely by baryons, and at the same time that part of the baryonic matter is not visible.

5.8 Dark Matter in Particles

If dark matter is necessary to explain a whole series of phenomena and if it is not composed of ordinary, or baryonic, matter, then what is it made of?

Already in the '70s, people began to speculate that dark matter was made up of some sort of particle. Several reasons pointed in that direction. To begin with, baryonic matter did not explain the estimated quantity of matter. Assuming that the universe is flat and therefore $\Omega = 1$, the estimates gave values for dark matter in the range between 20% and 30% (0.2–0.3 in terms of critical density) — much greater than that expected from the primordial nucleosynthesis, equal to about 5%.

The other fundamental aspect came from the study of the CMB. As already seen, after recombination, the plasma that made up the universe began to cool down and density perturbations began to grow due to gravitational instability resulting in structures. In order to allow time for structures, that is, galaxies, clusters, and superclusters to form, density fluctuations at recombination were estimated to be $\sim 10^{-3}$. The first studies on these density fluctuations yielded no results. Only in the late 1970s was it clear that these fluctuations had to be much lower than theoretical expectations. Then, astronomers began to think that dark matter could be made of particles. The first particles that were related to the missing mass were *neutrinos*.

In 1973 Cowsik and McClelland, after obtaining an upper limit for the neutrino mass (1972), first proposed the idea that some eV mass neutrinos could dominate the gravitational dynamics of galaxy clusters and the universe, and applied the result to the problem of the missing mass in the Coma cluster.

Before them, in 1966, Gershtein and Zeldovich discussed the role of *muon neutrinos*[3] in cosmology, but they did not

---

[3] Neutrinos are a family made up of three particles: the neutrinos associated with electrons, called *electron neutrinos* those associated with muons, called *muon neutrinos*, and those associated with the particle τ, called *tau neutrinos*.

connect them with the missing mass. They predicted the quantity of muon neutrinos in thermodynamic equilibrium in the primordial universe, the temperature at which the particles stopped self-annihilating, leaving a population of neutrinos that survived as a thermal fossil. They also predicted what their mass should be (<400 eV) and came to the conclusion that neutrinos of 10 eV would dominate the energy and mass of the universe. In the following years, the incorrect idea that neutrinos could help overcome the paradox of fluctuations in CMB was proposed. In a universe dominated by neutrinos, the growth of perturbations can begin earlier than in a universe made of baryons and, at recombination, have sufficient amplitude to form the structures.

The most significant representative of neutrino cosmology is the Belarusian physicist Yakov Zeldovich, an almost legendary character of Soviet physics and astrophysics, very versatile and prolific. He made significant contributions to fields such as nuclear physics, particle physics, relativity, astrophysics, cosmology, and material physics, and also played a crucial role in the development of the Soviet atomic bomb construction project. Among other things, in the 1960s, Zeldovich made significant contributions to the thermodynamics of black holes.

He connected accretion disks around black holes and quasars, and discussed primordial black holes. On Hawking's 1973 trip to Moscow, Zeldovich and Starobinsky showed Hawking that because of the uncertainty principle black holes had to emit and absorb particles. On the basis of these ideas, Hawking developed the well-known theory of black hole evaporation, dubbed *Hawking radiation*.

Even if it seems strange, black holes are not completely black. The phenomenon of evaporation can be simplified in this way. In the vacuum around the black hole, virtual particles and antiparticles appear and disappear. Because of the enormous gravitational field of the black hole, one of the particles in the pair may fall into the black hole, leaving the other particle free outside. This is equivalent to the creation of particles of mass equivalent to the mass of the particle left outside the black hole, at the expense of the energy of the black hole itself. Consequently, the black hole will gradually lose mass, in times proportional to the cube of its mass. For example, a black hole of a solar mass takes 10^{67} years to evaporate. Its temperature will also increase inversely proportional to its mass.

The formation model of Zeldovich's cosmic structures was based on neutrinos, which are particles with relativistic[4] velocities, and for this reason the model was called the *hot dark matter model*.

However, there is a non-trivial problem. In a universe dominated by neutrinos, their high-speed motion damps small-scale perturbations, leaving only the largest perturbations. In this model, the first structures forming are large objects such as galaxy clusters, called *pancakes*. Galaxies are formed later by fragmentation. This model is also called the *top-down model* because the formation of structures begins at large scales (some megaparsecs) and extends to small ones. It was soon clear that the neutrino-based model was not working. In fact, in the universe, galaxies are formed first and clusters later, contrary to the predictions of the HDM model. This discrepancy, together with others, such as the comparison with the simulations, led to the rejection of the model, and it was clear that neutrinos could not be the entire dark matter of the universe, and that it had to be made up with other types of particles. Neutrinos came back into vogue much later, in 1993, when Dodelson and Widrow proposed a fourth type of neutrino, called *sterile neutrino*, capable of explaining the entire dark matter. This hypothetical neutrino can only interact with other particles through gravity and through the mechanism of *neutrino oscillations*, with other neutrinos that allow a neutrino of a certain type to transform itself into other types of neutrinos.

In 1982, when it was clear that the neutrino dark matter model was not exact, several physicists, including Peebles, proposed that dark matter consists of other particles such as the *axion*, the *gravitino*, and the *photino*. Since these particles are not relativistic, the model was called the *cold dark matter*

---

[4] Particles with speed close to that of light.

model (abbreviated as CDM). Furthermore, since small-scale structures are formed before large-scale structures, these models are called *bottom-up models*.

Since these particles are not relativistic, they do not suppress small-scale perturbations. Thus, structures are formed from small to large scales, in agreement with the observations.

To summarize, dark matter cannot be made of baryons not only because it cannot explain the quantity of the missing matter, but because the baryon perturbations produced are much larger than those observed in the CMB. Similarly, it cannot be made of neutrinos, because they behaves like hot dark matter, and because they have in addition a very small mass. Despite being numerically very abundant,[5] they fail to make up 26% of the critical density. Instead, the CDM model predicts perturbations in accordance with those observed in the CMB and does not present the problems of the HDM model.

The question that returns in an iterative and annoying way is, what is dark matter made of? We think we know it must be made of particles, but in nature there are different types of particles. What is the right particle? The most trivial idea is to go and look for the particles we know of, those of the *standard model of particles*. For this, it is necessary to make a small excursion into this model.

---

[5] To get an idea, a surface of one square centimeter is crossed by 100 billion neutrinos in a second.

Chapter 6

THE GOLDEN AGE OF PARTICLE PHYSICS: THE STANDARD MODEL

Tyger Tyger, burning bright,
In the forests of the night;
What immortal hand or eye,
Could frame thy fearful symmetry?

— William Blake

The world around us seems to be a *continuum*, that is, objects can be divided indefinitely. Thales, Anaximander, Anaximenes, Empedocles had built a special conception of elements based on the idea of the continuum. Thanks to Aristotle, these ideas arose to the rank of philosophy and he imagined that the cosmos was made up of only four elements: earth, water, air, and fire, to which a fifth element called ether was later added. All earthly things were obtained from the combination of the first four elements, while the Sun and the stars were made from the fifth element. This erroneous conception dominated Western culture for millennia. There was already another worldview, supported by Leucippus and Democritus and their followers, the so-called *atomist theory*.

This theory was based on the idea that matter is made up of non-divisible basic elements, hence the term *atom*: indivisible. In contradiction with the so-called Aristotelian *horror vacui* (theory according to which nature abhors a void), the cosmos, according to the atomists, was constituted by an empty and unlimited space within which atoms moved relentlessly and, not subject to supernatural interventions, in a random motion. Their combination gave rise to material bodies. But what is there between one atom and another? According to Democritus, there is nothing, only empty space. In his view, there was a clear evidence of the existence of the atom when somebody cut an apple. Assuming that the atom is truly indivisible, a knife can cut an apple only if it passes in the empty spaces between the atoms, an idea similar to the modern one of how matter is made.

Unlike modern science, the ideas of atomists were only philosophy because they were not based on quantitative measurements or experiments. It took more than two thousand years for these ideas to rise to higher levels. Dalton's atomic hypothesis led to the first modern atomic theory of matter. Several other scientists attempted to sort the elements into groups with similar characteristics: Dobereiner with his triad theory, Cannizzaro, Chancourtois, and Newlands with the law of octaves. Then came the turn of Dmitri Mendeleyev, who in 1869 proposed a periodic system of elements that showed the regular variation of the chemical-physical properties when the elements were ordered according to increasing values of the atomic number. The regularity was so clear that Mendeleyev left empty spaces in his *table of the elements* to be inserted after their discovery. He even described the characteristics of those elements that were discovered a few years later, and were named *scandium*, *germanium*, and *gallium*.

In spite of everything, these were only clues to the existence of atoms. In the 20th century, several scientists, such as Ernst Mach, considered it an abstruse question, without the possibility of being verified. Twenty years later, in 1905, Einstein, with the interpretation of *Brownian motion*,[1] indirectly showed the existence of atoms.

It took about a hundred years and a great deal of theoretical and experimental effort to arrive at the formulation of a complete theory that described the fundamental interactions and particles connected to them, the so-called *standard model*

---

[1] That is, the random motion of particles suspended in a fluid. These movements had been explained by Einstein assuming the existence of molecules in the fluid which, colliding with the suspended particles, caused their motion.

of particles. This model, based on quantum field theories, describes all fundamental particles and interactions, except gravity. As we shall see, the first of the field theories on which the model is built is *quantum electrodynamics*, which describes the interaction between light and electrons, and which was later incorporated into the *electroweak theory*, which unifies electromagnetic and nuclear forces. Field theory that describes strong nuclear forces, called *quantum chromodynamics*, was also added to the model. The last missing piece of the puzzle was the discovery of the Higgs boson in 2012, included to solve the problem of the absence of the mass of the particles in the standard model.

6.1 Fields and Interactions

As mentioned, we have a model dedicated to describing three of the four fundamental interactions and the elementary particles related to them: *the standard model of particles*. The model is basically a theory of quantum fields. What does it mean?

A field is a function that assigns a value to each point in space. To understand this, let's do an example. Each point of a room has a temperature, and the set of all these temperatures is the *temperature field* of the room. This particular type of field is called a *scalar field* because it associates a single value of some variable with every point in space. In the case of a scalar temperature field, every point in space is associated with a temperature (Fig. 6.1).

Similarly, on a windy day, each point in space will correspond to the speed and direction of the wind. This too is a field, named *vectorial field*, because at each point you need to know the three components of the vector in order to know "what wind is blowing" (Fig. 6.2).

However, these fields are not fundamental fields; they are air properties. The gravitational and electromagnetic fields are *fundamental fields*. An easy-to-view field is the *magnetic field*. Just try a well-known experiment: place some iron filings on

The Invisible Universe

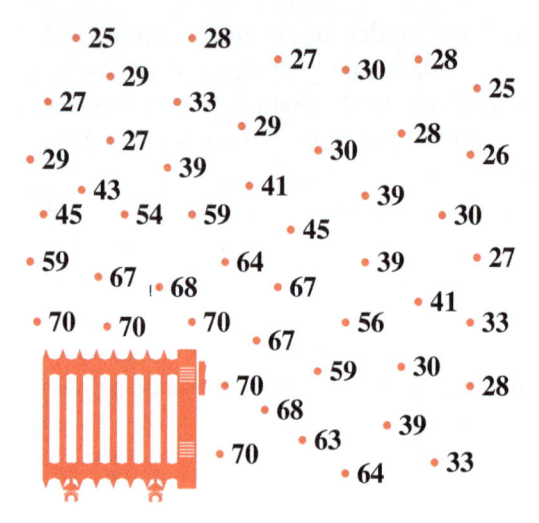

Fig. 6.1. Temperature field in a room. Each point has a temperature associated with it. Close to the stove in the room, the temperatures are higher than in the points that are far away.

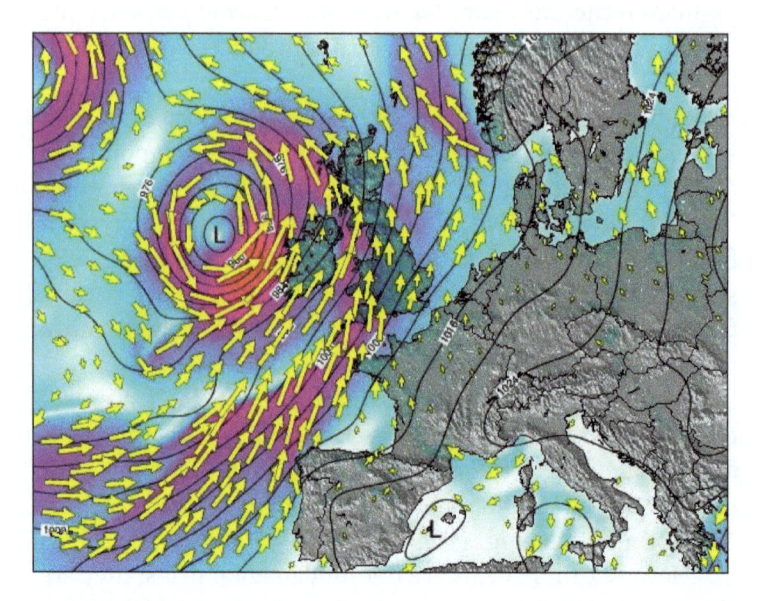

Fig. 6.2. Wind field in Europe. The yellow arrows indicate the wind direction and intensity (the larger the arrow, the stronger the wind). Credit: Tide-forecast.com.

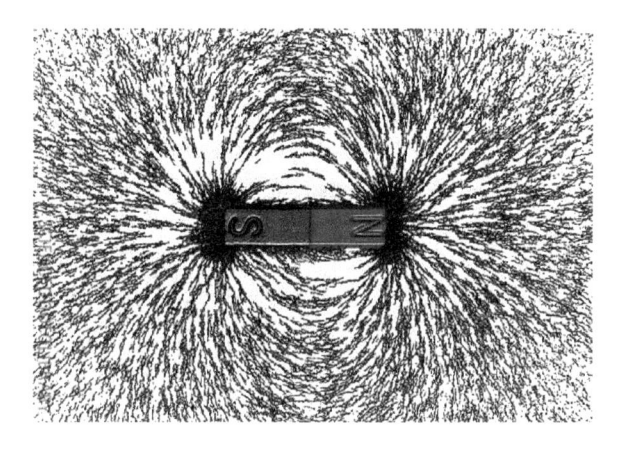

Fig. 6.3. Magnetic field lines of a magnet.

top of a surface and a magnet below it. The magnetic field lines will thus be clearly visible (Fig. 6.3).

Classical field theory was born with Michael Faraday. Faraday imagined that the lines seen in Fig. 6.3, called *lines of force*, were a sort of string whose tension was responsible for the forces of attraction and repulsion between the poles of the magnets. The field was for him a tangle of fine lines that fills the space in which we are immersed. With his idea, Faraday gave an answer to one of the passages in Newton's *Principia*:

> Gravity must be caused by an agent acting constantly according to certain laws, but whether this agent be material or immaterial is a question I have left to the consideration of my readers.

So Newton wrote the equations of gravity as a sort of instruction manual without bothering to explain how it actually worked. Faraday's idea finally explained how bodies manage to attract or repel each other without touching each other. This idea introduced for the electromagnetic field was later extended to the gravitational field by Einstein.

James Clerk Maxwell crystallized Faraday's insight into 20 equations, reduced to four by Oliver Heaviside.

Another fundamental field is the gravitational one. By combining Newtonian and field mechanics, the fact that the Moon continues to rotate around the Earth can be explained by noting that the Earth and the Moon are immersed in the gravitational field. The Earth acts on the gravitational field, or more precisely on the *gravitational potential*, which surrounds it. This acts on the "neighboring potential," and so on. In other words, the Moon will receive — not instantly, but at the speed of light — information about the presence of the Earth and will be attracted to it.

By applying quantum theory to fields, a *quantum field* is obtained, subject to the rules of quantum mechanics. Unlike classical fields, a quantum field does not associate a number with each point, but a function, an infinite number of points. In the case of quantum fields, if for example we measure energy in a room, it cannot take all the possible values available, but only multiple set values of a fundamental quantity. Furthermore, we cannot say that energy is associated with a certain point in the field, but we can only indicate the probability that a certain value of the energy is associated with a point. How many types of fields are there? A few dozen: the *neutron field*, the *photon field*, the *quark field*, the *neutrino field*, and so on. The fields associated with fermions are called *fermionic fields*, or fields of matter, and those associated with bosons, *bosonic fields*, or fields mediating interactions. Furthermore, the perturbations of a field can be transmitted to other fields.

In the view of modern physics, reality is made up of fields, entities that permeate space and manifest themselves through their vibrations. According to this view, particles and antiparticles are not fragments of matter, but packets of energy, or rather vibrations of quantum fields. By observing a field in detail, we see it "resolve" into particles. The energy of one field can transfer to another through interactions, a phenomenon present in processes in which a set of initial particles are transformed into another set of final particles. Fields are the real building blocks from which matter is made. Furthermore, they make interactions possible; that is, they are the fundamental forces of nature.

As the fields are extended entities in space, this explains several of the apparent paradoxes related to the behavior of subatomic particles.[2]

Particles are simple oscillations of fields. Like every daughter, every particle has its own father, a particular field. For example, the particles called *photons* are daughters of the *electromagnetic field*. A very interesting thing is the fact that for quantum mechanics, in our world, empty space is not completely empty, as we imagine it, but is made up of fields.

6.2 True and False Vacuum

Quantum vacuum is a sort of sea full of particles and antiparticles, *the virtual particles*, which are born and annihilated quickly. This continuous birth and death and the fluctuations of the vacuum are regulated, by the Heisenberg uncertainty principle. Therefore, owing to quantum uncertainty, small quantities of energy may appear, in the form of particle — antiparticle pairs, provided that they disappear in a very short time. As a consequence, each environment must present some quantum irregularities. Each field has

- an energy
- *real particles*, identified as excitations of the field itself — for example, an electromagnetic field has photons as associated particles
- *virtual particles*, that is, fluctuations in the vacuum that can disappear and appear continuously
- a vacuum state

Having said that, what is a vacuum?

Nature is "lazy," if we may say so, and therefore in physics every system tends to reach the state of minimum energy. Each field has an energy, and the minimum value of this energy is

---

[2] I refer, for example, to the delocalization and interference initially attributed to waves associated with the particles, and explain otherwise incomprehensible experiments such as Young's double-slit experiment.

called *vacuum*. Depending on the energy profile of the field, there can be a true vacuum or a false vacuum.

To better clarify the concept, we consider Fig. 6.4. The brown curve in Fig. 6.4 (left panel) represents the potential energy associated with the field. The horizontal axis represents the value of the field, ϕ, while the vertical axis gives its energy, $V(\phi)$. In Fig. 6.4 (left panel) the energy profile has the shape of a parabola and has only one minimum. In the minimum of the parabola, the field value is zero ($\phi = 0$) and the energy has its minimum possible value. This minimum represents by definition the *vacuum of the field*. Now let's again look at Fig. 6.4 (right panel). The point where the field is zero ($\phi = 0$) is not the minimum energy point, since there are two points (regions) with less energy, the two small "holes" on the left and right: on the left, the point with $\phi = -v$, and on the right the one with $\phi = +v$. These two points that are those of least energy are thus the *true vacuum* of this field. The point where the field is null ($\phi = 0$), not having minimum energy, is called *false vacuum*. This is an unstable condition for the field (the

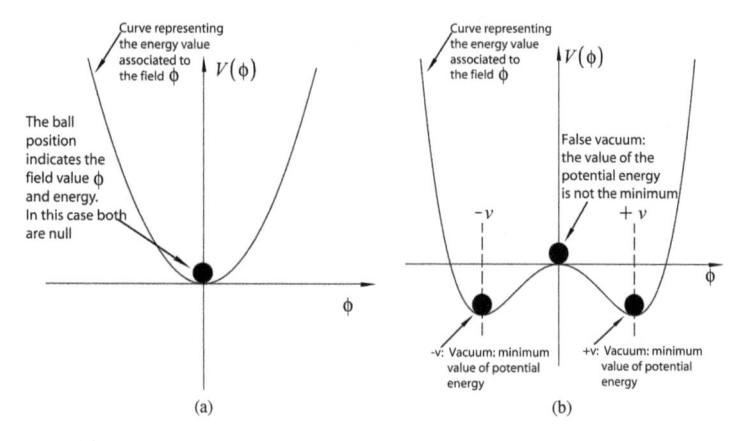

Fig. 6.4. (a) The potential energy, $V(\phi)$, of the field in terms of the value ϕ of the field. The field is in its minimum energy (black point) at the point where the field ϕ is zero. This point is the *vacuum of the field*, and no particles are associated with it. (b) Field with two energy minima in $\phi = +v$, $\phi = -v$, which are the points of *true vacuum*. The point with $\phi = 0$ is not the one with minimum energy, and for this reason it is called *false vacuum*. It can be seen that when the black particle is at the top of the hill, the field is in a false vacuum. So it falls into the minimum, that is, into the *true vacuum*.

black particle), and consequently it slides into the minimum energy, that is, into the *true vacuum*.

Let's put it in other words, trying to give physical meaning to the discussion. As mentioned, in general the minimum energy state of a field is called *vacuum*. The fields in the vacuum are subject to fluctuations around the minimum energy. These fluctuations are the virtual particles. If the vacuum of the field corresponds to its null configuration ($\phi = 0$), that is, if the value of the field is zero in the minimum energy (Fig. 6.4, left panel), the field has zero value in all space and will not contain real particles.[3] If we want real particles in the field, we need to supply it with energy and bring it to a configuration where its value is different from zero. By its nature, the field is "very quiet" in the least energy situation. This type of field (Fig. 6.4, left panel) has a true vacuum: when the field has a zero value, it is in the minimum energy, that is, in the vacuum state, and there are no real particles. In the case of the field in Fig. 6.4 (right panel), the null configuration ($\phi = 0$) does not correspond to the minimum energy state, and there are no real particles related to the field. In other words, if we don't want real particles associated with the field, we need to supply energy to the field. This kind of field by its nature is "hyperactive" and tends to fill the space with real particles unless we spend energy and "turn it off."

To summarize, fields with a *real vacuum* (Fig. 6.4, left panel), if not excited, have only small energetic excitations around the null value. Fields with *false vacuum* (Fig. 6.4, right panel) are always full of particles.

The field in Fig. 6.4 (left panel) is similar to the majority of fields, such as the electromagnetic one, while that in Fig. 6.4 (right panel) is similar to the *Higgs field*, as described in section 6.7.

Is it possible to verify that the vacuum with the defined characteristics really exists? The answer is yes. There are three effects that are explained thanks to the existence of the vacuum: the *Lamb effect*, the *Casimir effect* and the *Sauter-Schwinger effect*.

---

[3] However, virtual particles, fluctuations around the null value, are present.

In 1928 Dirac obtained an equation that brings together quantum mechanics and special relativity. In 1947 Willis Lamb and Robert Retherford found a tiny difference in energy between two levels (2s and 2p) of the hydrogen atom (different states for the orbit of the electron around the nucleus), which, according to Dirac's equation, should have had the same energy. This was a big problem since the equation had made very precise predictions and was the best equation they had to describe the structure of the atom. Dirac's equation does his job very well; the problem is related to the existence of vacuum. Owing to the fluctuations of the vacuum, particles and antiparticles appear and disappear at every point, in a period of time inversely proportional to the energies of the particles created. Photon pairs are also created. The electrons of the atom that are in two states (2s and 2p) have a high probability of being close to the nucleus. The fluctuations of the vacuum move them away from the proton, increasing their energy. Ultimately, the void is responsible for the cited energy discrepancy. This effect is called the Lamb effect, and it is of considerable importance because his explanation prompted the formulation of *quantum electrodynamics.*

In 1948, Hendrik Casimir showed that two parallel metal plates separated by a small distance in a vacuum undergo an attractive force (Fig. 6.5). The force originates because the vacuum structure is modified by the presence of conductive plates.

Simplifying, the plates are contained in a vacuum. The wavelength of the virtual photons that are located outside the plates is not limited by their presence, while the virtual particles between the plates must have a wavelength that is an integer sub-multiple of the distance between the plates.

So, outside the two plates, more fluctuations occur than between the plates, yielding a deficit of energy balance between the plates that tends to bring them closer.

Is there a method to "observe" quantum fluctuations? In theory, yes, and it's called the *Sauter–Schwinger effect*, proposed in 1931 by Fritz Sauter. In 1951 Julian Schwinger gave a complete theoretical description based on works by

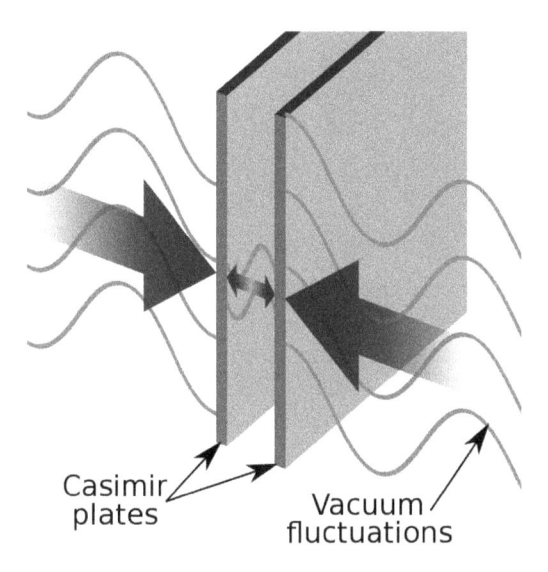

Fig. 6.5. Casimir effect. Only fluctuations with a wavelength equal to an integer sub-multiple of the distance between the plates are allowed between the plates. Outside, the wavelength of virtual particles does not have this limit, so that a force is produced that tends to bring the plates closer together. Credit: Wikipedia, Davide Mauro.

Heisenberg and Hans Euler. In a vacuum, pairs of particles and antiparticles are created which are readily reabsorbed by the vacuum. If we apply a powerful enough electric field, the pairs could be separated (e.g., electron–positron pairs). To do so, however, huge fields are needed, and so far this has not been possible.

Another key point related to fields is the fact that forces, or interactions, arise from deformations of fields in space and are related to symmetries. So, in addition to being the building blocks of the world, fields are the origin of interactions, more commonly called forces. To explain it, we need to link the fields with the concept of symmetry.

6.3 From Symmetries to Reality

Symmetry permeates our existence. We have an intuitive idea of it. For example, it is immediate to note that the butterfly on

Fig. 6.6. Bilateral symmetry in a butterfly, and an asymmetrical sofa.

the left of Fig. 6.6 has a bilateral symmetry, while the sofa on the right is not symmetrical.

For example, if we look at a sphere from different directions, it will always appear the same to us, because it is *invariant* in a rotation. Similarly, if we rotate a cylinder with respect to its main axis, it will always seem the same to us. So a physical system has symmetry if its properties do not change after a transformation performed on it. There are several types of symmetry, such as *continuous* and *discrete*. If we rotate a vase with a circular base relative to the main axis, its appearance will not change whatever the angle of rotation. This is an example of *continuous symmetry* (Fig. 6.7, left panel). If the vase has a hexagonal base, it will have the same appearance only for rotations of 60 degrees, and we have a *discrete symmetry* (Fig. 6.7, centre). If, at this point, we imagine to infinitely multiply the vases, as in Fig. 6.7, right, in order to have one in every point of space, we will have many more possible symmetries. If we rotate all the vases to the same angle, we will have performed a *global transformation*, while if we rotate each vase to a different angle, we will have carried out a *local* or *gauge transformation* (Fig. 6.7, right panel).

Symmetries can also break. In this case, we speak of *spontaneous symmetry breaking*. The term seems very abstruse and has few relations with reality. However, it is a very frequent process that, on certain occasions, we have all been witnesses to. We can make several examples. In everyday life, we would certainly have seen the transformation of water into ice, or the transformation of ice into water. These are examples of spontaneous symmetry breaking. Let's consider some water

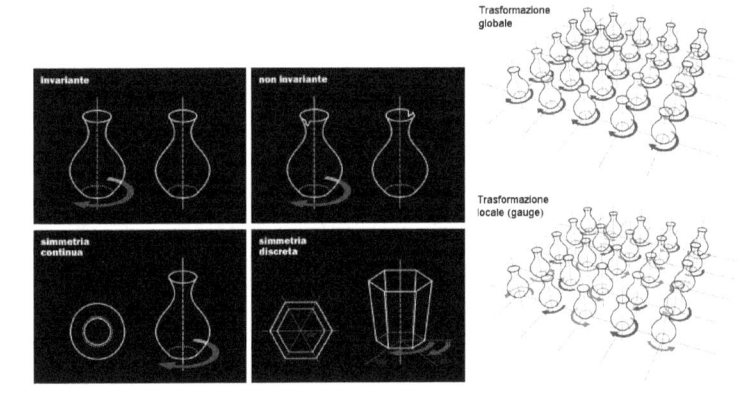

Fig. 6.7. Various forms of symmetry. Credit: Asimmetrie/INFN/F.Cuicchio.

in a container and let it cool, putting it in the refrigerator. At the beginning there will be water in the vase, which is a symmetrical structure: from whichever direction we look at it, it will always appear the same. Over time, the first ice crystals will begin to form and then the water will freeze. At this point, the symmetry of the water will have been spontaneously broken, since ice is a crystalline structure with privileged directions. Another example is that of a thin liquid layer, for example oil in a tray. If we heat it from below, we will go from a situation in which the oil is distributed evenly to that in which hexagonal cells form on the surface.[4] The system went from a higher symmetry (homogeneous distribution in the pan) to the presence of the hexagonal cells. To conclude, we have shown another example of spontaneous symmetry breaking in Fig. 6.4b. Initially, the field, represented by a sphere, is at its maximum. The system is symmetric by rotation. The situation with the sphere at the top is unstable and it tends to fall downwards, that is, towards the minimum. When this happens, the system loses rotational symmetry.

In summary, spontaneous *symmetry breaking* is a mode of loss of symmetry following a symmetry transformation. In the

---

[4] The hexagonal cells that are formed are called Rayleigh-Bénard cells, named after the experimenters who observed them.

transformation, the system changes its structure and naturally loses its symmetry, while the underlying laws and equations still maintain it. Sometimes we speak of *hidden symmetry* to highlight that symmetry is not lost at a fundamental level.

It is interesting to note that the phenomenon of spontaneous symmetry breaking explains a large amount of symmetrical patterns found in nature such as the striped body of zebras and tigers, as pointed out by Turing in 1956, and that of jaguars, as discussed by Maiani in 1990.

In addition to the symmetries present in the real world, there are others in the equations that regulate the dynamics of physical systems. In 1918 Emmy Noether showed that for every continuous symmetry there is a *conservation law* and a corresponding conserved quantity. For example, the invariance by translation in space gives rise to the *conservation of the momentum*, the product of mass by speed, while a translation over time gives rise to the *conservation of energy*.

The continuous symmetries of which we have spoken refer to transformations of the space-time coordinates. However, other continuous symmetries may exist in the case of transformation of fields, *internal symmetries*.

The link between symmetries and knowledge of the laws of nature reached its peak with the advent of quantum mechanics. It was shown that the gauge symmetry, already mentioned, can generate the fundamental forces and the particles that transmit these interactions. José Ignacio Illana, in *Le particelle elementari: il cuore della materia* (*The Elementary Particles: The Heart of Matter*[5]) gives a very interesting and intuitive description of the meaning of internal symmetry, of gauge, and how forces can be generated from symmetries. We describe it below.

Let's consider a river that flows in a certain landscape. The landscape of the region is a field because every point in the region corresponds to a height, which determines the dynamics of the river. If we raise the landscape to a height h, by applying a *global transformation*, the path of the river will

---

[5] *Le particelle elementari, il cuore della materia* (RBA Italia).

not be changed. So there is a symmetry. The river will continue to flow undisturbed and the situation will be the same as before. We can say that "the equations that describe the river are invariant in case of global transformation of the field." The flow of the river remains constant and represents the *quantity conserved*. Now suppose we carry out a local transformation of the field by raising the landscape to a height h different from point to point. As a consequence, the river will no longer be able to follow its usual course, and "the equations will no longer be invariant." The water will come out of its riverbed, and the flow will no longer be the same. That is, the quantity that was previously preserved is not anymore. To maintain symmetry in the presence of local transformations, a "compensating force" (an *interaction field*, also called *gauge field* or *connection field*) must be added to ensure that the river re-enters its natural (original) riverbed and resumes its journey. Obviously, the required field of interaction is determined by the symmetry that is desired to be maintained. These symmetries are referred to as *gauge symmetries*.

So the origin of the fundamental interactions is a manifestation of the gauge symmetries of the universe: the symmetries regulate the interactions.

6.4 Work in Progress: Building the Standard Model

Summarizing the adventure and the discoveries that led to the formulation of the standard model of particles is not easy, but there were fundamental moments that remained in the history of physics, and mark the path of the story.

One of these moments was the identification of the electron, an elementary particle having dimensions less than 10^{-22} m and mass about 10^{-30} kg (0.511 MeV) (see Box 1, Unit of Temperature and Energy), by J.P. Thomson in 1897 and the discovery of radioactivity by Henri Bequerel in 1896. In 1898–99, Rutherford identified two types of radiation called α and β, and Paul Villard identified a third type of radiation called γ. The study of β radiation, a few decades later, would have provided clues on new particles and forces of nature. The discovery of radioactivity put an end to the idea of the indestructibility of atoms and a few decades later would have provided ideas on new particles and forces of nature. This allowed Rutherford in 1911 to determine the distribution of the charge in the atom and the discovery of the positive charge at its center, called *protons*.

The hypothesis of quantization of radiation by Planck, to explain the oddities of the black body emission, allowed Einstein to explain an important effect of physics, the *photoelectric effect*, attributing particle characteristics to radiation. Gilbert Lewis coined the term *photon*, or packet of light, responsible for electromagnetic interaction. The same hypothesis was used by Bohr to explain atomic stability by assuming that the electrons moved only on particular orbits in which the electron does not emit or absorb energy, allowing the spectral lines to be explained.

In 1924, de Broglie pointed out that particles are associated with the physical properties of waves, which led to the idea of *wave–particle dualism*, that is, the fact that these entities cannot be described only as waves or particles: in certain experiments the wave can behave like a particle, and in others a particle can behave like a wave.

With the joint effort of a generation of physicists, a theory was developed that explained, in the case of systems with speeds lower than that of light, what happened in the atomic world. That theory is the quantum mechanics of Heisenberg and Schrödinger, built thanks to the contribution of a host of physicists, including Planck, de Broglie, Bohr, Born, Dirac, and so on. Also, on the origin of quantum mechanics in Heisenberg's version, there is one of the many anecdotes that fill physics, such as that of the oscillations of the candelabrum in the cathedral of Pisa leading to the discovery of the law of isochronism of the pendulum (Galileo), or that of the fall of the apple and Newton's intuition that the gravitational force acting on Earth was the same that moved the planets. According to this anecdote, told by Carlo Rovelli, Heisenberg, strolling in the park near the Theoretical Physics Institute in Copenhagen, saw for a moment a man pass under a lamppost and then disappear. Then he saw him reappear when he passed under another lamppost. It was obvious that an object as large as a man does not have the property of appearing and disappearing, but in a microcosm this could happen. An electron could appear and disappear, make "quantum jumps" between different orbits in an atom. In a microcosm, the trajectory may no longer be a sensible concept. It is possible that an electron exists only when you look at it, that is, when it interacts with a photon. From this episode, and many calculations, the first formulation of quantum mechanics was born, the Heisenberg version. On the basis of de Broglie's hypothesis, Schrödinger's theory introduced the *wave equation* whose unknown, dubbed *wave function*, is related to the probability of finding a particle in a given spatial region. The advent of this theory, devoid of the strange formalism of Heisenberg's, was received with great praise. Even Einstein claimed that he was happy with the discovery, and that Heisenberg's theory was surely wrong.

Among other things, this theory changed the classic view of the Rutherford and Bohr atom that had been represented in a similar way to a planetary system. In the new theory, the electron was distributed in space around the nucleus as a wave of probability in regions called *orbitals*. In 1925 Uhlenbeck and Goudsmit showed that the atomic spectrum needs electrons to *spin*, which can be interpreted in a simplified, not very correct, way as an intrinsic rotation around the particle's axis.[6] The spin allows to classify the particles in fermions, having a

---

[6] The non-correctness of the idea was already known since the beginning. The history of the discovery is somehow funny. As Goudmith told it (https://www.lorentz.leidenuniv.nl/history/spin/goudsmit.html), after Uhlenbeck and Goudsmit wrote the paper related to spin, and delivered it to their supervisor, Ehrenfest, Uhlenbeck started to have doubts and went to speak to Lorentz, who said, "Yes, that is very difficult because it causes the self energy of the electron to be wrong." Then Uhlenbeck went to speak to Ehrenfest about the paper and said, "Don't send it off, because it probably is wrong; it is impossible, one cannot have an electron that rotates at such high speed and has the right moment." Ehrenfest replied, "It is too late, I have sent it off already.... Well, that is a nice idea, though it may be wrong. But you don't yet have a reputation, so you have nothing to lose."

half-integer spin (1/2 $\hbar$, 3/2 $\hbar$, 5/2 $\hbar$ ...) and bosons having integer spin (0 $\hbar$, 1 $\hbar$, 2 $\hbar$ ...).

In 1928, Dirac obtained an equation that described the properties of fermions in the relativistic regime, which led him to hypothesize the existence of an anti-fermion for each fermion, of opposite charge and identical mass. The postulated electron antiparticle was discovered in cosmic rays by C. Anderson in 1932 and was called the *positron*. In the same year, Chadwick discovered the *neutron*, hypothesized starting from the conservation of energy in nuclear reactions. The neutron has a mass of about 939 MeV, a little larger than that of the proton, 938 MeV, and size $\hbar$ of the order of 10^{-15} m.

6.5 The Flirtation between Light and Electrons

Among the many contributions that Dirac made to physics, there is also a first formulation of a quantum theory describing the interaction between matter and energy. The calculation method used is called *perturbation theory*, based on a series of successive approximations. The method had been introduced in celestial mechanics to calculate the orbits of the planets. As a first approximation, it is assumed that the motion of a planet is not influenced by that of the others, then the perturbations due to the other planets are introduced. In other words, one starts from a simpler problem and then adds more complicated "pieces." For example, to determine the orbit of the Earth considering the dominant gravitational effect, and that of the Sun, the contributions of the other planets are added. Using an analogy, the method works like this. Suppose you want to know the price of your antique watch studded with diamonds. If you ask a watchmaker, he will take a look at it and give you a first estimate. You point out that he didn't consider the jewels. This would produce a second, more precise estimate. If you then tell him that it is an ancient and rare piece, the watchmaker will still correct his estimate. In other words, the final value is reached with a series of corrections and iterations. However, the perturbation method has limitations.

If we were to calculate the orbit of three planets of very similar mass, the method would be practically unusable, because there is no predominant effect that gives an estimate to "refine" with subsequent calculations. In this case, other methods are needed, "non-perturbative" methods, or the use of a computer.

This method, together with those of classical mechanics, was essential for the emergence of quantum mechanics, in particular in the works of Born and Heisenberg. The method gave correct values up to a few percent. Trying to get more precise estimates incurred the appearance infinite values. The solution to the problem was given after the Second World War by Sin-Itiro Tomonaga, Julian Schwinger, Richard Feynman, and Freeman Dyson, who introduced a set of techniques for dealing with the problem of infinities, called *renormalization*.

The resulting theory, a quantum theory of the electromagnetic field, the first theory of quantum fields, was called *quantum electrodynamics* (QED). It describes all phenomena involving charged particles interacting by means of electromagnetic force, while including the theory of special relativity. In this theory, charged particles interact with each other through the exchange of virtual photons, which live for so little time that they cannot be detected. For example, the interaction between two electrons that repel each other by exchanging a photon can be imagined as a situation in which two people exchange a heavy ball, as shown in Fig. 6.8. The boat of the person throwing the ball recoils; that is, it moves backwards. The boat of the person receiving the ball will also move. This is similar to what happens in the microscopic world when two particles exchange a particle.

In short, the interaction is explained in terms of the exchange of virtual particles. Moreover, the range at which an interaction is felt is inversely proportional to the mass of the mediating particles. Charged particles interact by exchanging virtual photons; quarks interact by exchanging another form of photons, *virtual gluons*. Such particles that transfer interactions are generally referred to as *gauge bosons*. The heavier the gauge boson, the shorter the interaction range, and vice versa. This rule is not valid for the theory of strong

Fig. 6.8. Analogy between the exchange of a heavy ball between two persons on two boats and the exchange of a virtual particle between two particles. Elementary forces, at a subatomic level, appear when one matter particle emits a virtual particle. Similarly, in our everyday experience, when somebody throws a heavy ball off a boat, the boat recoils, and another person on another boat receiving the ball moves away from the other boat.

interactions in which gluons, although having zero mass, are confined within the nucleus.

Thus, for example, the electromagnetic interaction mediated by the photon that has zero mass will have infinite interaction radius. The weak nuclear interaction mediated by the W and Z bosons with masses, respectively, 80.39 and 91.19 GeV (i.e., about 90 and 100 times the mass of a proton) is short-range.

In theory, both matter and interaction are described by fields, and in addition to describing experimental data with unprecedented precision, the theory represents the starting point for explaining other interactions in nature, except gravity.

In addition to gravity and electromagnetism, there are two other forces in nature that act on very small distances, of the order of the size of the atomic nucleus. The first is called *strong nuclear force* and acts on quarks. The second is the *weak nuclear force* responsible for the decay of particles such as the neutron. It allows the sun to shine, given that solar energy comes from the conversion of protons into helium and the transformation of protons into neutrons. QED is the most successful theory of physics. Richard Feynman pointed this out by making an analogy: if you measured the distance between New York and

Los Angeles with the precision of QED, this distance would be known at the level of the size of a hair. After the success of QED, the goal of the theorists was to study these two forces.

6.6 The Strange Story of Weak and Electroweak Interaction

Further developments came from the study of β radiation discovered by Rutherford, a process now known as β decay, that is, the decay of a proton into a neutron, a positron, and a neutrino, in which electrons and neutrinos are emitted by the nucleus, which has a positive charge. Rutherford erroneously suggested that the nuclei consist of protons and electrons. Moreover, he thought that they coupled, forming a neutral particle. At that time, it was thought that only the electromagnetic interaction operated in the nucleus, and a nucleus made only by positive charges would break. Furthermore, this hypothesis explained the β radiation, in which electrons are emitted by the nucleus. In 1932 Chadwick repeated an experiment conducted by Bothe and Becker in Berlin and by the Joliot-Curie spouses in Paris. By bombarding a beryllium plate with ionized helium nuclei, neutral particles were emitted with a mass, at that epoch, measured in 1.007 times greater than that of the proton. Chadwik concluded that the particle was the neutron hypothesized by Rutherford. Only in 1934 it was clear that the neutron was a new fundamental particle, and not the combination of a proton and an electron as Rutherford thought. More precise measurements give the *neutron* a mass of 939 MeV, a little larger than that of the proton, 938 MeV, and a size of the order of 10^{-15} m.

In 1934 Fermi explained the phenomenon as the conversion of a neutron into a proton with the emission of an electron from the nucleus. The explanation given by Fermi started from the idea that these particles were oscillations in quantum fields and that there was an interaction between them — a situation similar to that of wall clocks, which, placed close to one another, in a short time in some way, still not perfectly known, synchronize through the interaction of sound waves. So the electron and the neutrino do not appear from nothing, but the acceleration of the neutron field is transferred to that of the electron, the neutrino, and the proton.

However, there was a problem: in the decay the energy is not conserved. For this, Pauli hypothesized the existence of a particle that was baptized *neutrino*, by Amaldi, during a conversation with Fermi. Today, it is known that this particle has a very small mass, between 100,000 and 10,000,000 times less than that of the electron. Its interaction with matter is so weak that the neutrino was discovered after 22 years of wild research, in 1956, by Cowan and Reines. To understand how weakly the neutrino interacts with matter, just think of the fact that to block half of a beam of neutrinos would require a lead wall of a thickness larger than a light-year. Fermi's theory of weak interaction had to be revised to be re-normalizable like QED, and to take account of the so-called *parity symmetry*, which we mentioned in Chapter 2. We recall that in a *parity transformation* the spatial coordinates are all reversed: (x, y, z) becomes $(-x, -y, -z)$. Weak interactions "don't like" to respect this rule (more precisely, weak interactions are not invariant under the previous transformations). This means that weak interactions behave differently for a given system and for its reflection in a mirror.

In 1956 Schwinger proposed a *gauge theory* for weak interaction, and after the discovery of the particles W^+ and W^-, in 1983, he proposed that as the photon is the mediator of the electromagnetic force, so these particles, which are bosons, are the mediators of the weak force. In order to solve some technical problems, a third particle was added, the Z^0 boson, by Glashow and Salam.

Gauge symmetry does not allow particles to have mass, but in reality it is observed that the bosons W^+, W^-, Z^0, and quarks have mass. In order to solve this problem, one has to make sure that the symmetry is eliminated in some way, that is, a transition is necessary between the situation in which the particles have no mass to that in which they have it. Using a term coined by Baker and Glashow, one must have a *spontaneous breaking of symmetry*.

Unfortunately, there was another problem: it was believed that in every theory compatible with special relativity, massless particles called *Nambu–Goldstone bosons* had to emerge from each symmetry breaking. This idea was even put in a theorem form.

In fact, in 1961, Steven Weinberg visited the Imperial College, where he discussed the problem with Abdus Salam. They came to the conclusion that there was no way to avoid such particles. Their unfortunate conclusion gave rise to the so-called Goldstone theorem, published in 1962 by Goldstone, Salam, and Weinberg. The only physicist who realized that there was an exception to this rule was Philip Anderson. In 1963, he pointed out that a counterexample to *Goldstone's theorem* could be found in the laboratory, observing the spontaneous symmetry breaking in the oscillations of an electron plasma or the mass acquisition of photons in superconductors. In both cases, the Goldstone bosons appeared, but gained mass. However, this happened in a non-relativistic framework, and it was necessary to verify if this could happen in the theory of quantum fields.

The way to escape Goldstone's theorem and solve the problem was shown in 1964 by three groups: Francois Englert and Robert Brout in Brussels, Peter Higgs in Edinburgh, and two Americans (Gerald Guralnik and Carl Hagen) and an Englishman (Tom Kibble). In short, it was shown that although the Goldstone bosons made their appearance, they did not materialize because they were "absorbed," as we will see better in the following.

For various events, as often happens in physics, the mechanism to solve the problem was named after only one of the proposers: the *Higgs mechanism*.

After the formulation of the quoted mechanism, in 1967, Weinberg proposed a unified theory of weak and electromagnetic interaction and introduced the quoted mechanism in it. In 1971, Gerard 't Hooft proved that the theory was re-normalizable, that is, self-consistent.

6.7 The Higgs Mechanism

The particles of the standard model can be divided into two groups. On the one hand, there are the particles that measure the interactions, namely photons, W^+, W^-, Z^0, and gluons, and on the other, the particles of matter, namely quarks and leptons. The particles that have a mass are W^+, W^-, Z^0, quarks, and leptons.

Unfortunately, a problem that plagued the theory was the fact that the particles of the model had to have zero mass for reasons of symmetry. Some mechanism was necessary to break the symmetry, but in the symmetry breaking, the well-known massless Nambu–Goldstone bosons appeared. A sort of trick was needed to make them disappear or hide. This "trick" is the Higgs mechanism, in which Goldstone bosons are "eaten" by gauge bosons, which become massive. How does this happen?

As we know, particles have a spin, which is often described in classical mechanics and in early quantum theory as a rotation.[7] If we consider an electron, its spin is ½ and in measurements we can find only two *spin states*, or *polarization states*, or *degrees of freedom*, which represent the ways in which a field can vibrate: ½ and –½. Each polarization state can be considered as a different particle or a different degree of freedom. *A fundamental point to remember is that massive particles have three degrees of freedom.*

For a particle with spin 1, we can have two possibilities: if the particle is massive (e.g., in the case of W and Z bosons), we could find the values –1, 0, 1. The particle has three degrees of freedom. This means that the fields of massive particles can have an up-down vibration (and vice versa), a left-right vibration (and vice versa), and forward-backward vibration (and vice versa). If the particle has no mass, as in the case of photons, we

---

[7] As already reported, the notion of spin as rotation around an axis is incorrect. This quantity was not foreseen by non-relativistic quantum mechanics; it had been introduced as a result of several experiments. It is automatically obtained in the relativistic version of quantum mechanics, by the Dirac equation, and does not correspond to a rotation around an axis.

would find −1 and +1, that is, only two degrees of freedom.[8] So the fields of massless particles can vibrate only in two ways: up-down (and vice versa) and from right to left (and vice versa). These oscillations are shown in Appendix C (Fig. C1).

To summarize, a scalar field, with spin 0, has only one degree of freedom. Vector fields with spin 1 have two degrees of freedom if they are massless and have excitations perpendicular to the direction of propagation (1, −1). If they have mass, they have three degrees of freedom, with two perpendicular excitations (1, −1) and one longitudinal (0) (see Appendix C).

So how do the particles W^+, W^- and Z^0 acquire their mass?

Before the symmetry breaking, the electroweak interaction was mediated by four fields: the fields B, W_1, W_2, and W_3 (Fig. 6.9) and the corresponding bosons were all massless.

We can consider them as the parents of the fields of the photon γ and those of W^+, W^-, and Z^0. Consequently, they have only two degrees of freedom, excitations perpendicular to the direction of propagation, and zip at the speed of light. To make them massive, one needs a third degree of freedom, the longitudinal one (forward-backward oscillation). We need three scalar particles that, combining with W_1, W_2, W_3, provide each with the missing longitudinal degree of freedom. These scalar particles, the *Goldstone bosons*, are produced at the end of the electroweak era, when the universe had 10^{-12} s and temperature was 10^{15} K in the electroweak symmetry breaking phase mentioned in Chapter 2. As described in Appendix C, the Higgs potential underwent a transformation. The field "slipped" from the false vacuum to the true vacuum, like a ball that descends from the top of a hill to the valley (see Fig. C3).

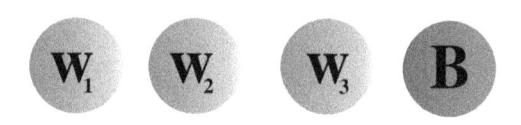

Fig. 6.9. The electroweak interaction mediators.

---

[8] It should be noted that if 0 is found, this does not mean that the particle has no spin.

In this event, four scalar particles (see Fig. 6.10) were generated which we refer to as H_1, H_2, H_3, and H, or also as H^0, H^+, H^-, and h three of which (H^0, H^+, H^-) are Goldstone bosons, eager to help W_1, W_2, W_3 to become massive. H_1, H_2, H_3, alias H^0, H^+, H^-, provide the degree of longitudinal freedom needed by W_1, W_2, W_3 to become massive.

After W_1, W_2, W_3, absorb H_1, H_2, H_3, that is, the missing degree of freedom, immediately become massive. In Figs. 6.9–6.13, we show some cartoons describing how the bosons W_1, W_2, W_3, by eating H^0, H^+, H^-, transform into W^+, W^-, and Z^0. W_1 and W_2 combine and, eating H^+, give rise to W^+ (Fig. 6.11).

Similarly, W_1 and W_2 combine and eating H^-, give rise to W^- (Fig. 6.12).

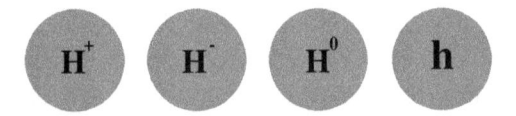

Fig. 6.10. The three massless Goldstone bosons H^+, H^-, H^0, and the (massive) Higgs boson, h.

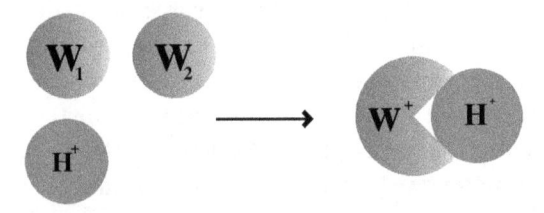

Fig. 6.11. Generation of W^+ from the combination of W_1 and W_2 and eating of H^+. Credit: Modification of work by Flip Tanedo, Quantum Diaries.

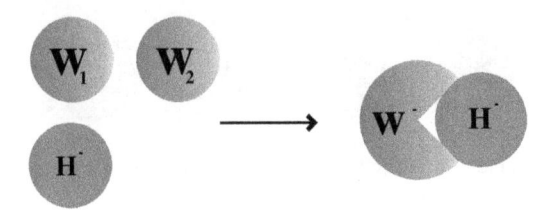

Fig. 6.12. Generation of W^- from the combination of W_1 and W_2 and eating of H^-. Credit: Modification of work by Flip Tanedo, Quantum Diaries.

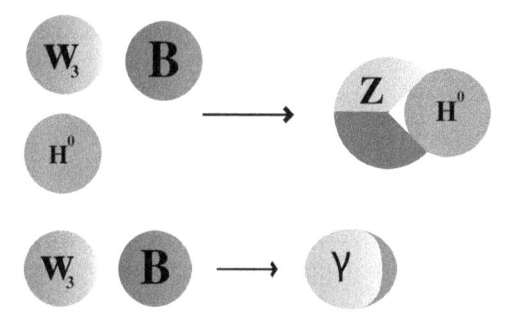

Fig. 6.13. Top: Generation of Z^0 from the combination of W_3 and B and eating of H^0. Bottom: Generation of the photon, γ, from the combination of W_3 and B. Credit: Modification of work by Flip Tanedo, Quantum diaries.

The boson Z is obtained when W_3 and B combine and eat H^0. W_3 and B also combine (Fig. 6.13). No Goldstone boson is left, and then the photon (combination of W_3 and B) remains massless, and h, which is *the Higgs boson* discovered in 2012, remains free.

In Appendix C, a more precise description of the Higgs mechanism is given.

In the case of fermions (quarks and leptons), mass acquisition occurs in a different way, with the so-called extension of the Higgs mechanism to *Yukawa's interaction*, an interaction that couples the fermion field with that of the Higgs. Intuitively, in their motion these particles "feel" the presence of the Higgs field, which fills the whole space, and begin to interact with it. When fermions move in space, they interact with the Higgs field. Using a not very correct, but effective, analogy, the field of Higgs slows them down like honey does with an object that moves in it. Entangled in the Higgs field, the particles find it more difficult to move, as if they had mass and their speed becomes less than that of light. Particles such as electrons assume mass when the Higgs boson "intertwines" the two polarizations of the electron. The operation requires energy, supplied by the Higgs field, which is perceived as the resting mass of the particle. The mass grows as the intensity of the interaction grows.

If we could disable the Higgs field, we would go back to the initial state with four massless bosons. The quarks would lose their mass, the two electron components with different polarization would decouple and lose their mass, and the atoms and matter would be annihilated.

A legitimate question is, why the other fields cannot do what the Higgs field does? What distinguishes the Higgs field from other fields, for example from the electromagnetic one? This difference was described in more detail in section 6.2 when we talked about fields. Unlike the other fields, the Higgs field is always different from zero in its state of minimum activity, $\phi = 0$, as seen in Fig. 6.4 (right panel), and Fig. 6.14. It even occurs that the Higgs field has more energy when it is null, that is, when $\phi = 0$, than when it is not null. Contrary to the usual fields, if we want the Higgs field not to invade space with its bosons, we must supply energy. This is the difference that distinguishes it from other fields. Particles, sensitive to it, always have a certain probability of interacting with it and become massive.

The intuitive description of the Higgs mechanism is due to David Miller, a physicist from University College London. He won the prize offered by Minister William Waldegrave (a bottle of champagne) to the physicist who had managed to explain in an intuitive and exhaustive way how the mechanism works and what it was used to find out.

Miller imagined a salon full of people evenly distributed in space, chatting among themselves as a VIP arrives. As the VIP

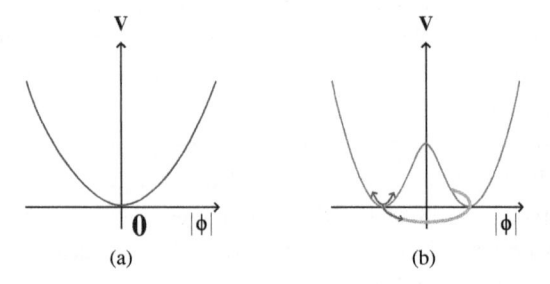

Fig. 6.14. The potential V of a field like the electromagnetic one (left) and that of the Higgs field (right). When the field "falls down" from the hill to the minimum, it can have an up and down oscillation generating the boson H and, on the circumference of the minimum, generating the bosons H_1, H_2, H_3.

moves, he attracts people and a crowd forms. The crowding produces resistance to the movement of the VIP, and it looks as if the VIP is acquiring mass. A particle moving in the Higgs field faces a similar situation.

The concept of the Higgs boson is explained by imagining that someone is circulating a voice in the lounge. Those who listened to it first will report it to the neighbors. The room will be crossed by a wave of "crowds of people" that, like the VIP, "gain mass." The Higgs boson would resemble this crowding.

It is often said that the Higgs boson gives mass to objects. This is not exactly correct. The Higgs mechanism provides mass to only the elementary particles. An important point to note is that the mass of any object, such as a table, is due in minimal part to the Higgs field. Everyday objects are made up of protons and neutrons.

Each proton is made up of two up-quarks, with mass 2.3 MeV, and one down-quark, with mass 4.8 Mev. So if the mass of a proton were all due to quarks, it would have to weigh 2.3 + 2.3 + 4.8 = 9.4 MeV. In reality, the proton has a much larger mass, 938 MeV. This means that the Higgs field that provides mass to the quarks contributes only about 1% to the mass of the proton. The mass of the proton is basically due to the fluctuations of the vacuum (Fig. 6.15). The quarks that

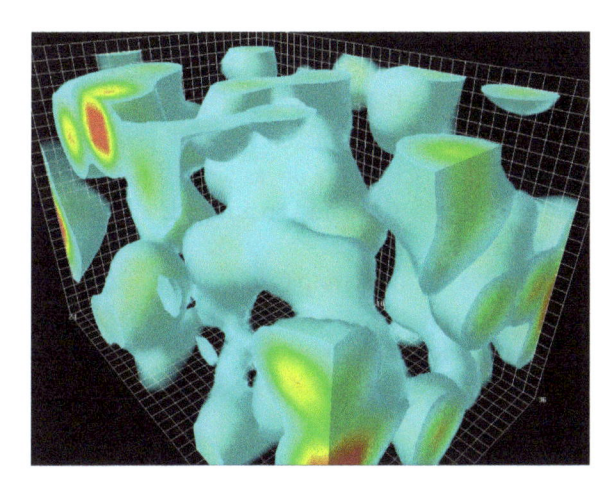

Fig. 6.15. Fluctuations of the vacuum in a proton. Credit: Wikipedia, Derek B. Leinweber.

make up the proton are confined in the proton, within which the quantum vacuum is subject to fluctuations that create and destroy quarks and gluons continuously and in a very short time. The interaction of the quarks that make up the proton with the fluctuations of the vacuum generates energy (see the Casimir, Lamb, and Schwinger effect in Chapter 6).

If the Higgs field did not exist, the quarks and leptons would have no mass and would move in space at the speed of light. The consequence is that protons, neutrons, nuclei, and atoms would not form the universe and, finally, us.

So far, we have talked about the theory of how elementary particles gain mass. Is there evidence that this is really the case? The answer is yes. A large part (3/4) of the Higgs mechanism was discovered in 1983 at CERN by Carlo Rubbia and Simon van der Meer, with the discovery of the W^+, W^-, and Z^0 bosons. The cherry on the cake, or the Higgs boson, was discovered on July 4, 2012, thus completing the construction of the standard model.

6.8 The Dynamics of Colors and the Atomic Nucleus

Meanwhile, the study of cosmic rays had allowed to identify several particles. In 1936 Carl Anderson, the positron discoverer, together with Seth Neddermeyer, discovered a negatively charged particle, like the electron, but about 200 times heavier. It was called μ *meson*.

A meson is a particle made up, as we will see, of two other elementary particles: a quark and an antiquark. These particles "feel" the strong nuclear force.

The reception of the new particle was not the warmest, since it was not clear why there should have been a heavier cousin than the electron. Physicists were so stunned that Isaac Rabi learned of the discovery and exclaimed, "And who ordered it?"

Moreover, physicists had not understood that in reality the particle was not a meson but precisely a cousin of the electron, about 200 times heavier. It belongs, together with neutrinos, to the family of leptons that "feel" the weak nuclear force. In the 1970s, history repeated itself and another cousin of the electron and muon was discovered, the *tau*, or *tau meson*, having a mass approximately 3,500 times that of the electron. Only in 1947 were the mesons predicted by Yukawa really discovered: the *pion*, or π *meson*, and the *kaon*, or *k meson*. To study the deep structure of matter, in the 1950s, particle accelerators began to be built. Accelerators allow to produce collisions with energies much greater than those of the particles involved in the collision, thus producing particles of greater mass. According to the equation $E = mc^2$, for a particle to be generated it is necessary that the energy be greater than its mass at rest. As the accelerator energies increased, the number of particles produced became so large that there was talk of a *zoo of particles*.

Murray Gell-Mann and Yuval Ne'eman showed that the particles could be arranged in a two-dimensional scheme that indicated a wider but also more approximate symmetry, called "the way of the octet," since they remembered the geometric structure of the octets.

The new particles belonged to the hadron family. Physicists began to think that these particles were not really elementary, and in 1964 Gell-Mann and Zweig proposed the *quark model*, confirmed in 1968 with experiments at the Stanford Linear Accelerator (SLAC), distinguishing the properties of the quarks with the terms *up* (u), *down* (d), and *strange* (s). The model described the proton as consisting of the three *uud* quarks and the neutron constituted by the *udd* combination. Particles such as pions are composed of a quark and an antiquark. However, the model had a flaw. The particle (baryon) Δ^{++} consisted of three up-quarks, and this violated the Pauli exclusion principle. The solution proposed by Han, Nambu, and Greenberg was to assume that there is a new quantum number for quarks called color. It was later recognized that it was a sort of charge (similar to the notion of electric charge, but at the same time with various differences with it), called *color*. So each of the quarks had three different colors, and suddenly the number of quarks tripled. Han and Nambu suggested that the strong interaction was mediated by eight gauge bosons, called *gluons*. These are the ideas behind the strong interaction theory, *quantum chromodynamics*, which takes its name from color, in Greek *chróma*. Gross, Wilczek, and Politzer showed that the color force, unlike electromagnetism, weakened with distance, an effect called *asymptotic freedom*. In addition, the color strength increases when quarks are trying to move away, resulting in the inability to see free quarks. This phenomenon is called *confinement*. Today, in addition to the up, down, and strange quarks, we know that there are three other quarks: *charm* (c), *bottom* (b) (or beauty), and *top* (t) (or truth). The quarks are combined into three pairs — (u, d), (c, s), (t, b) — of increasing mass, and each with the same pair of charges, 2/3 e^-, −1/3 e^-. The quarks therefore have three generations of particles and similarly the leptons: (e^-, v_e), (μ, v_μ), (τ, v_τ). Quarks were discovered between the '70s and the '90s. The last and heaviest of the quarks, the top, was observed at Fermilab in 1995.

6.9 Summary: Particles and Interactions of the Standard Model

6.9.1 *Particles*

The fundamental building blocks that make up the world are the fields. However, their oscillations manifest themselves as particles. The standard model contains two large families of particles: fermions and bosons.

Fermions take their name from Enrico Fermi and obey the *Pauli exclusion principle*. They follow the *Fermi–Dirac statistic* and therefore have a half-integer spin (1/2 $\hbar$, 3/2 $\hbar$, 5/2 $\hbar$ …) and are the constituents of matter. There are two types of fermions: the quarks, constituting the protons and neutrons, and the leptons, constituting electrons, muons, tauons, neutrinos, and their respective antiparticles.

The *bosons* take their name from Satyendranath Bose, they follow the *Bose-Einstein statistic* and from the *spin-statistic theorem*, they have integer spin (0 $\hbar$, 1 $\hbar$, 2 $\hbar$…).

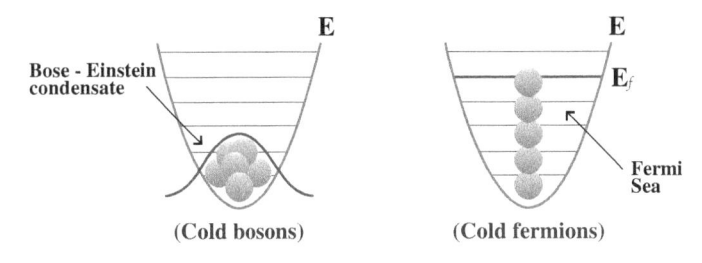

```
                           E                              E
Bose - Einstein  _____                    _____
condensate  \   _____                    _____ E_f
             \  _____                    _____
              \ _____                ↙   _____
                _____/                    _____/  Fermi
                                                            Sea
              (Cold bosons)                  (Cold fermions)
```

Fig. 6.16.   Bosons and fermions.
Source: Modification of work by Fred Bellaiche.

Fermions, following the Pauli exclusion principle, occupy much more space than bosons, as shown in Fig. 6.16. On the left side is represented a gas of bosons and on the right one of fermions at temperatures close to absolute zero (−273.16 degrees centigrade). The boson gas collapses to form a *Bose–Einstein condensate*. Fermions cannot place themselves in a state that takes up little space. They arrange to occupy energetic states of increasing energy up to a state of maximum energy called *Fermi energy*. This is why fermions are the constituents of matter, while bosons are the vectors of interactions

Bosons are distinguished in *gauge bosons* or vectors that mediate the forces and in *mesons*, stable particles consisting of a quark and its antiparticle, antiquark. Particles composed of a large number of particles, such as protons and neutrons, and atomic nuclei can behave like bosons or fermions depending on the total spin. Another classification of the particles according to the forces to which they are subjected or their mass is as follows:

1. *Hadrons* (from the Greek "strong") are particles subject to strong nuclear force and can be divided into

    - *Barions* (from the Greek "heavy"): They are fermions. They consist of nuclear particles, the baryons, such as protons, neutrons, the particle Λ, *Sigma*, Δ, Ξ, Ω (and their respective antiparticles), and heavier and non-nuclear particles, *hyperons*. Baryons are not elementary particles, but made up of *quarks*. There are six types of quarks: *up* (u), *down* (d), *strange* (s),

The Invisible Universe

Standard Model of Elementary Particles

	three generations of matter (fermions)			interactions / force carriers (bosons)	
	I	II	III		

QUARKS

- ≈2.2 MeV/c² ; 2/3 ; ½ — **u** up
- ≈1.28 GeV/c² ; 2/3 ; ½ — **c** charm
- ≈173.1 GeV/c² ; 2/3 ; ½ — **t** top
- 0 ; 0 ; 1 — **g** gluon
- ≈124.97 GeV/c² ; 0 ; 0 — **H** higgs

- ≈4.7 MeV/c² ; −1/3 ; ½ — **d** down
- ≈96 MeV/c² ; −1/3 ; ½ — **s** strange
- ≈4.18 GeV/c² ; −1/3 ; ½ — **b** bottom
- 0 ; 0 ; 1 — **γ** photon

LEPTONS

- ≈0.511 MeV/c² ; −1 ; ½ — **e** electron
- ≈105.66 MeV/c² ; −1 ; ½ — **μ** muon
- ≈1.7768 GeV/c² ; −1 ; ½ — **τ** tau
- ≈91.19 GeV/c² ; 0 ; 1 — **Z** Z boson

- <1.0 eV/c² ; 0 ; ½ — **ν_e** electron neutrino
- <0.17 MeV/c² ; 0 ; ½ — **ν_μ** muon neutrino
- <18.2 MeV/c² ; 0 ; ½ — **ν_τ** tau neutrino
- ≈80.39 GeV/c² ; ±1 ; 1 — **W** W boson

GAUGE BOSONS / VECTOR BOSONS

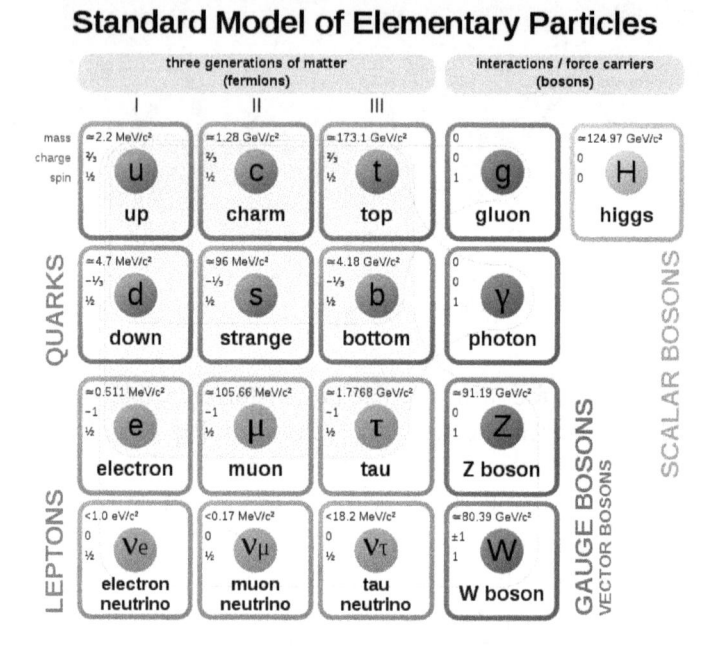

SCALAR BOSONS

Fig. 6.17. Particles and interactions in the standard model. Credit: Wikipedia.

charm (c), *bottom* (b) (or *beauty*), and *top* (t) (or *truth*). The quarks are combined in three pairs — (u, d), (c, s), (t, b) — of increasing mass, and each with the same pair of charges 2/3 e⁻, −1/3 e⁻ (where e⁻ is the electron charge). Each quark has *three colors*, so we have 18 quarks. Protons consist of two up- (u) and one down-quark (d): (uud). Neutrons from one up- and two down-quarks (udd). Protons have a mass of about 938 MeV and neutrons are slightly more massive, 939 MeV, and have a size of the order of 10^{-15} m. Outside the nucleus, neutrons are unstable and have an average life of approximately 15 minutes. Protons are almost immortal particles, given that their average life is greater than 10^{34} years, much greater than the age of the universe. The average life of the proton is important because some Grand Unification Theories (GUTs), theories that unify strong, weak, and electromagnetic nuclear forces, require that the

proton decay. The baryons are assigned the so-called *baryonic number*, which is $B = +1$ for baryons and $B = -1$ for anti-baryons. For the other particles, mesons and fermions, $B = 0$. The baryon number is an almost always preserved quantity, that is, the baryon number before a reaction is the same as the one after.
- *Mesons* (from the Greek "middle"): They are bosons subject to a strong nuclear force with intermediate masses between the baryons and the leptons. They consist of a quark and an antiquark.
- *Leptons* (from the Greek "light"). They have no strong interaction, are elementary particles, and are fermions with spin 1/2. Examples of this family are charged particles such as electrons with a mass of about 10^{-30} kg (0.511 MeV) and dimensions less than 10^{-22} m, *muons* with a mass 207 times that of the electron, *tauons* with a mass about 3500 times larger than that of electrons, and their respective antiparticles. The family also contains neutral particles, the *neutrinos*. Neutrinos have a very small mass, between 10,000 and 1 million times smaller than that of the electron. They come in three flavors: *electron, muon*, and *tau neutrino* and can oscillate, that is, transform into each other (*neutrino oscillations*). Like quarks, leptons have three generations of particles (e, νe), (μ, νμ), (τ, ν$_\tau$). Like baryons, leptons are assigned a leptonic number: $L = +1$ for particles, $L = -1$ for antiparticles, and $L = 0$ for hadrons.

Ultimately, we have three families of leptons, six particles, and three of baryons, another six particles. To each of the particles corresponds an antiparticle of equal mass and opposite charge.

A very interesting point is that matter is made of fermions and only of particles of the first family, namely quarks u and d and electrons. A legitimate question at this point is, why are there two other families of particles if they do not constitute matter, and why is there a mass hierarchy? This and many other questions, as we will see at the end of this section, show the limitations of the *standard model*.

Another discrepancy of the model is that neutrinos have no mass, while the phenomenon of neutrino oscillation, that is, the transformation of one of the three types of neutrinos into another type, implies that they have mass, however small. One of the possible solutions, better described in Appendix D, is that there is another neutrino, the *sterile neutrino*, which interacts with other particles only with gravity and can interact with other neutrinos also with the mechanism of neutrino oscillations. The *seesaw mechanism* (Fig. 6.18) ensures that the mass of the neutrino is inversely proportional to that of the sterile neutrino. If this latter has a mass of the order of 10^{16} GeV, the neutrino has a mass of about 10^{-2} eV, which is the expected mass for neutrinos.

This mechanism is based on the existence of new physics at the scale of the Grand Unification Theory, 10^{15}–10^{16} GeV.

6.9.2 Interactions

In nature, there are four interactions or forces: *gravitational*, *electromagnetic*, *strong nuclear*, and *weak nuclear force*.

The standard model only describes the last three.

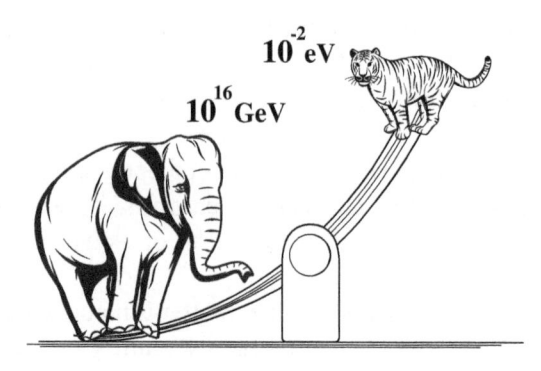

Fig. 6.18. The seesaw mechanism. The larger the mass of the right-handed, sterile, neutrino (see Appendix D), the smaller the mass of the usual neutrino.

- *Electromagnetic force*: It can be attractive and repulsive and decays with the square of the distance. It is responsible for electromagnetic phenomena (emission and absorption of radiation) and chemical reactions. It holds electrons around nuclei. The electromagnetic force is mediated by the *photon*.
- *Strong nuclear force*: It is the most intense. It is responsible for the existence of the atomic nucleus, has a range of action in the order of the size of the proton, and grows as the distance decreases. It is mediated by eight bosons, the *gluons*.
- *Weak nuclear force*: It is responsible for radioactive decay and contributes to making the stars shine. It has three mediators, the bosons W^+, W^-, Z^0, which, unlike the mediators of the other forces, have mass, respectively 90 and 100 times heavier than that of the proton. Their mass comes directly from the Higgs mechanism. The weak and electromagnetic forces are manifestations of a single force, the *electroweak force*.
- *Gravitational force*: Although not described by the standard model, it is of fundamental importance. It has a never-observed theoretical mediator, the *graviton*. It is the weakest of forces, 10^{25} times less intense than the weak one, 10^{36} than the electromagnetic one, and 10^{38} than the strong one, but it dominates in large spaces and in the universe. It is always attractive and responsible for the formation of structures and the rotation of planets around stars, galaxies, and so on.

The *Higgs boson*, discovered in 2012, supplies mass to elementary particles such as quarks and W^+, W^-, Z^0 bosons. Its measured mass is 125 GeV, much smaller than the theoretical expected value. This problem, which we discuss in the next chapter, is usually indicated with the *hierarchy problem*, and its solution implies the existence of physics beyond the standard model.

The 12 elementary particles (not counting quark color, which would bring them to 24) do not feel the various forces in the same way. All "feel" the gravitational force, while only the charged ones perceive the electromagnetic force. Strong force is felt only by quarks.

Protons, neutrons, and electrons form *atoms*. The simplest one, the hydrogen atom, is made of a proton and an electron. A gold atom has 79 protons and 118 neutrons, and for electrical neutrality, 79 electrons. An atom is basically an empty structure in quantum mechanics.[9] To get an idea, if the core had the diameter of a tennis ball (7 cm in diameter), the closest electron would be at a distance of 3 km.

Although the standard model is a very successful theory with incredible precision, it is not a complete theory, because it is not able to fully explain the nature of the world. Several open questions remain. For example, why is the world made only by a generation of quarks and leptons but in nature there are three of them? Are quarks and leptons really elementary particles? The mass of the particles is not given by the model but is obtained experimentally. Is there any way to get it from theory? That is, why does the standard model not foresee the mass of the particles? And again, the universe contains only matter, and if so, why? How can gravitation be integrated into the model? Is it possible to build a theory in which all forces are unified? How can we solve the hierarchy problem (we will discuss this issue in Chapter 7).

Finally, the standard model does not provide the existence of dark matter and energy. Other theories besides the standard model, which we will illustrate later, have been proposed that introduce models for dark matter and dark energy. However, they have never been confirmed experimentally.

---

[9] In quantum field theory, this is no longer the case. The space between electrons and protons contains virtual particles.

Chapter 7

DARK MATTER CANDIDATES

What we know is a drop, what we don't know is an ocean.

— Isaac Newton

7.1 Dark Matter Identikit

Dark matter and its identification have many similarities with a detective movie. It is very unlikely that investigators will be at the crime scene when a crime is committed. Investigators, if lucky, can find clues like fingerprints, hair torn during a scuffle, footsteps, and the like. Each clue gives information about the culprit. Having gathered all the clues, an identikit of the culprit can be traced. With painstaking work, it is hoped that the culprit will eventually be caught. In short, in the words of Sherlock Holmes, "when you have eliminated the impossible, whatever remains, however improbable, must be the truth."

So we need to compose an identikit of dark matter if we want to find it, and if we are dealing with particles, we need to understand what kind of particles they are, what range of masses they must have, what kind of interactions they "feel."

We said at the beginning that baryons and neutrinos cannot be all the missing matter in the universe. Following Sherlock Holmes's methods, we can try to see if any of the particles of the standard model are right for us. In order to consider whether a possible dark matter particle exists in the standard model, it is necessary to consider what the characteristics of the dark matter particles must be.

- Dark matter is neutral, otherwise it would interact much more than predicted by the observations. However, there is a scenario for dark matter in which it can have a tiny charge.
- Dark matter must have weak interaction with ordinary matter, that is, it feels weak interaction. The levels of interaction with ordinary matter are given to us, for example, by the impact of clusters of galaxies such as the *Bullet Cluster*. It is subject to gravitational force. The interaction through the other forces, if they exist, must be very weak, otherwise it would have already been observed.
- Dark matter cannot be ordinary matter because the limits coming from *primordial nucleosynthesis* theory and from the CMB indicate that ordinary matter is <5%, much less than the expected quantity of dark matter. Furthermore, baryonic matter cannot form structures as we observe them today.
- Dark matter must be stable. In the early universe, the amount of dark matter was almost equal to that present today.
- Dark matter interacts very little with itself. One of the various tests is the Bullet Cluster observations that we discussed, which allow estimates of the degree of self-interaction.
- Dark matter must be cold, that is, at decoupling it must have a velocity dispersion much smaller than that of light. If it were not, like neutrinos, primordial perturbations on small scales would be smoothed or deleted from their motion.
- The dark matter particle must be so abundant as to account for 26% of the mass expected by observations and theories.
- Finally, even if dark matter is fundamental for the formation of cosmic structures, it must not influence the results of primordial nucleosynthesis. Although it is fundamental for the formation and evolution of galaxies, it must not modify stellar evolution.

We can now use the previous criteria to see if in the standard model there are particles that fulfill the previous conditions. All the particles of the standard model feel gravity, so from this point of view they are all fine. The condition that the particles are stable excludes the quarks s, b, c, t, the muon, μ, the tau, τ, the gluons, the particles Z and W. For example, the μ decays in 2 millionths of a second in a τ, which decays even faster. The condition that dark matter is not hot (with relativistic velocity dispersion) excludes the photon, γ, and the neutrinos v_e, v_μ, v_τ. The condition that they are not baryonic excludes the quarks u, d, and also the electron, e^-. No more particles remain in the standard model, that is, dark matter particles must come from physics beyond the standard model or from particles other than those shown in Fig. 6.17. There is a subtle exception: the case of the *axion*, which is a candidate of dark matter and is linked to the standard model.

Despite this debacle, physicists have not given up trying to find new dark matter candidates. The imagination of physicists has been so prolific that in recent decades it has been realized that it is not difficult to propose candidates for dark matter. Indeed too many have been proposed. All the particles proposed as dark matter candidates, except the axion, are particles that have nothing to do with the standard model. As early as 1977, Steven Weinberg and Benjamin Lee published an article on neutrino-like but more massive particles, the *heavy leptons*. This study stimulated several others in which particle physicists began to apply their knowledge to cosmology. Interest poured into theories that extend the standard model. We will now see the best known.

7.2 A Special Symmetry: Supersymmetry

The interest of physicists was directed to a theory that extends the standard model called *supersymmetry* (SUSY). In this theory, each particle corresponds to its supersymmetric particle, or s-particle.

In the case of bosons, the supersymmetric companions are fermions, and vice versa. To give some examples, the photon corresponds to the *photino*, the gluon to the *gluino*, the Z particle to the *zino*, the W particle to the *wino*, the graviton to the *gravitino*, and the Higgs particle to the *higgsino*. The supersymmetric companions of the fermions are bosons, which are indicated by the name of the corresponding fermion preceded by an s. For example, the electron corresponds to the *selectron*, the neutrino to the *sneutrino*, and the quarks to the *squark*.

Ordinary and supersymmetric particles have different spins. For example, an electron has spin ½, while a *selectron* 0.

Why was supersymmetry introduced?

Some of the reasons are related to unsolved problems of the standard model, as we will now discuss.

We saw in the previous chapter that although the standard model is a successful theory, making frighteningly precise predictions in some cases, it has limitations. One of the limits mentioned is the *hierarchy problem*, which can be defined as the enormous difference between the expected value for the mass of the Higgs boson (or of the Higgs field), of the order of the *Planck mass*, 10^{19} GeV, and that measured, 125 GeV.

When we try to estimate a quantity (e.g., the mass of a particle) in the standard model, we proceed by iterations using *perturbation theory*. To arrive at high precision, it is necessary to calculate higher-order quantum corrections, taking into account the different contributions of various types of virtual particles. In the macroscopic world, a ball is an object with a well-defined shape and size. If we want to measure its weight, we just use a scale. If, on the other hand, we have an electron, a microscopic charged particle, it does not go around by itself, but always has a swarm of suitors, the virtual particles. If we want to estimate its mass, we have to "dress" it with the various types of virtual charges that contribute to its mass, adding one contribution after another. That is, the mass of the electron is given by its *naked mass*, considering it not charged, plus the mass of the particles that cover it. Since the electron is a point particle, if we do not consider the effects of virtual particles, its energy would be infinite. In the case of the electron, the screening by virtual particles makes a sort of miracle, eliminating the infinite. The same does not happen for the Higgs boson.

To calculate the mass of the Higgs boson, we will do the same thing, taking into account the contributions of the virtual particles.

These contributions are generally very large and help to increase the mass value up to the upper limit of Planck's mass, 10^{19} GeV, which represents the energy at which the effects of quantum gravity become important. Looking in more detail, the contributions of virtual particles to the mass of the Higgs boson come from fermionic and bosonic fields (particles), which contribute with an opposite sign to the final result.

One might therefore think that these effects cancel randomly. In reality, we know that very large positive or negative sums of magnitude usually result in large positive or negative quantities. It may happen that these sums are canceled by accident, but this is very unlikely. When the sum of many contributions (numbers), much larger than the measured value of a quantity, give a smaller result than the contributions, we speak of a *fine-tuning problem*. In order for these sums to result in small numbers, something needs to change the result. The situation is similar to that of a player on the stock exchange who for a certain period buys and sells shares for millions of euros and in the end finds himself holding a few cents. For this to happen, the operator must intervene in some way, adjusting the operations until they lead to the desired result. So we are left with a dilemma: Why is the mass of the Higgs boson much smaller than the contributions? Why is there a *hierarchy* between the order of magnitude of the mass of the Higgs boson and the Planck mass that separates them?

To solve the problem, we can introduce something that acts as the stock exchange operator and that can make ends meet. This something is *supersymmetry*.

Supersymmetry introduces fermionic and bosonic fields that correspond exactly, and the final effect of the sums can be that of a "miraculous" cancellation giving rise to a mass of the Higgs boson much lower than the Planck mass and equal to what has been observed.

Supersymmetry also solves another problem: the unification of fundamental forces. The intensity of the elementary interactions is not constant, but depends on the energy. For example, the intensity of the electromagnetic interaction on Earth and that in the center of a giant star are different. The same happens with weak and strong interactions. In the world we live in, the electromagnetic and weak force are different forces. However, at high energies, they become a single interaction, the electroweak one, with energies of the order of 246 GeV. Extrapolating the idea, Georgi and Glashow came to the conclusion that, at higher energies, the electroweak force unifies with the strong one, giving rise to a single force. As already seen in the history of the universe, this phase in which electromagnetism, the strong and weak interaction, constituted a single *super-force* mediated by the hypothetical bosons X and Y, is called *grand unification*, and *the grand unification theory* foresees that the three fundamental forces, excluding gravity, become one at energies of the order of 10^{15}–10^{16} GeV. Calculating the variation of the coupling in terms of energy or temperature, it can be seen that there is actually a trend towards unification with the growth of energy. This seems to go hand in hand with the expectation that when the universe was very young and very hot, there was only one interaction.

However, if you do the calculation in the framework of the standard model, you see that there is no unification, as seen in the left panel of Fig. 7.1: the three straight lines do not meet. As discussed by Dimopoulos, Raby, and Wilczek in 1981, reworking the calculations with supersymmetric models shows that the unification takes place (Fig. 7.1, right panel).

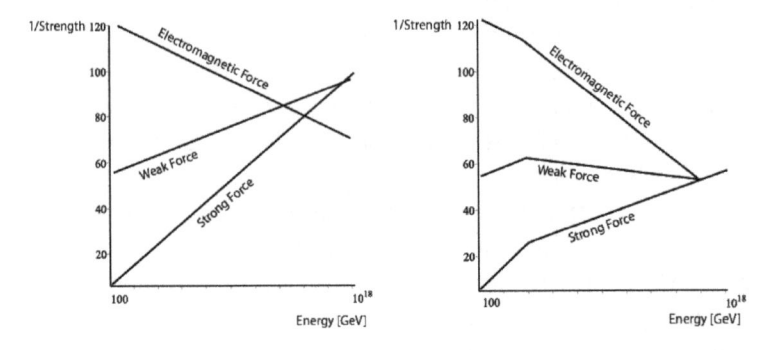

Fig. 7.1. Unification of forces without (left) and with (right) supersymmetry.

It is highly improbable that this result is due to chance. However, one of the GUT forecasts could be tested, namely the *non-conservation of the baryon number* and the *decay of the proton*. According to the theory, the proton should have an average life of 10^{31} years. Several experiments have tried to verify this prediction with a negative result, and furthermore the refinements of the theoretical calculations have raised the average life of the proton to at least 10^{34}–10^{35} years (see section 10.2 for more details).

A third interesting aspect of supersymmetry is that it comes with a bonus: some of its particles have the right characteristics to make up dark matter.

In the standard model, the strong and the weak nuclear force and electromagnetism are mediated by gauge bosons. Using the same ideas, the existence of a boson, mediator of gravitational interaction, called *graviton*, was hypothesized. This particle has never been observed, and moreover the theory which should predict it, *quantum gravity*, has not yet been constructed. Supersymmetry provides for the existence of a companion of the graviton, a fermion called *gravitino*. Supersymmetry offers other possible candidates for dark matter,

In 1982, although supersymmetry was still in its infancy, Heinz Pagels and Joel Primack proposed that this particle could solve the problem of missing mass. In 1981 Savas Dimopoulos and Howard Georgi laid the foundations for a simplified supersymmetry model, called the *minimal supersymmetric model* (MSSM), which considers the minimum number of particles and fields that achieve supersymmetry. Although simplified, the model has 120 new parameters that must be entered arbitrarily, giving the freedom to build specific supersymmetric models.

The MSSM allowed to open the way for new supersymmetric particles. In 1983 Haim Goldberg and in 1984 John Ellis, together with four collaborators, proposed that dark matter consists of a new particle called *neutralino*. In the MSSM the "mixture" of the photon superpartners,[1]

---

[1] A superpartner is the supersymmetric particle corresponding to a particle of the standard model. For example, the *photino* is the photon superpartner.

namely the Z particles, and two neutral Higgs bosons form 4 neutralinos, the lightest of which is stable.

Did we mention two Higgs bosons? Yes, in supersymmetry, there are 5 Higgs bosons, 3 neutrals and 2 charged (H^+, H^-, and 3 H^0 bosons).

Since it was proposed for decades, the neutralino has become the most popular and studied dark matter candidate. Just think that about 50% of the articles on dark matter candidates are on neutralino-related particles. This particle is part of a larger family, the *weakly interacting massive particles* (WIMPs), that is, particles subject to weak interaction. In order for a particle to rise to the rank of dark matter candidate, it must meet the requirements already seen. The neutralino has some important virtues:

- it is neutral
- it has weak interaction with ordinary matter
- it is not baryonic matter
- it is cold dark matter having dispersion of velocities much smaller than that of light.

Two other requirements remain to be verified: stability and how much they contribute to the matter that makes up the universe.

The neutralino is a supersymmetric particle, and in supersymmetry, as in GUT theory, there are interactions that violate the conservation of the number of baryons and leptons. This leads to a rapid proton decay, in a few years. Since the proton has an average life much longer than that of the universe, constraints must be imposed on the theory so that there is no conflict with the observational data.

For this reason, a new symmetry was introduced in supersymmetry, called *R-parity*, linked to the spin of the particles and to the difference between baryonic and leptonic numbers, conserved quantities. It has a value of 1 for the particles of the standard model and −1 for the supersymmetric particles. Preservation of the R-parity requires that supersymmetric particles be created and destroyed in pairs, leaving at least one

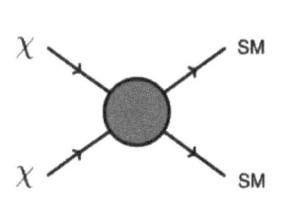

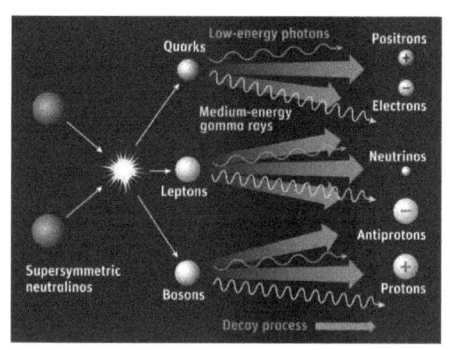

Fig. 7.2. Left: The way in which dark matter particles can produce the observed dark matter abundance. Dark matter particles annihilate into particles of the standard model. χ stands for dark matter particle, and SM for standard model particle. Right: The various particles of the standard model in which dark matter particles can decay are shown. Credit: Sky & Telescope/Gregg Dinderman (https://www.universetoday.com/127971/weighing-dark-matter/).

supersymmetric particle (the lightest) stable on cosmological times. In other words, a heavy supersymmetric particle can decay into a lighter one of the standard model (Fig. 7.2), but the lighter supersymmetric particle cannot decay.

The masses of neutralinos range from 10 to 10,000 times the mass of the proton. This range of masses is important to satisfy another characteristic required of dark matter particles: to have an abundance in accordance with the observations.

One question that has no clear answer is how dark matter was produced. As we will see, there are many other candidates besides the neutralino, and the WIMP, and the formation changes according to the type of particle. There are several theories that try to answer this question. In the case of WIMPs, the process that leads to a *freeze-out* is the following.

- In the very hot primordial universe, it is thought that WIMPS were in thermal equilibrium with ordinary matter. This means that the WIMPs interacted, giving rise to standard model particles, and the reverse process also happened (Fig. 7.2).

- Two competing processes fix the cosmic quantity of WIMPs:
  1. *Universe cooling.* While at high temperature the process of annihilation of WIMPs was equilibrated by the production of these particles due to annihilation of standard matter particles, at lower temperature the WIMPs annihilation was more important than the reverse process (Fig. 7.2). Consequently, the number of WIMPs decreased. The annihilation processes make dark matter decrease exponentially.
  2. *Universe expansion.* Since the universe expands, over time, it will be increasingly difficult for a WIMP to find another one to annihilate with.

     The condition in which WIMPs do no longer annihilate is reached, and their quantity *freezes out*. They will survive until today, leaving a *fossil quantity*.

The final abundance of WIMPs is controlled by their mass and interaction strength, more precisely by the *annihilation cross section*.

Particles with masses in the range 10–10,000 times the mass of the proton and with an interaction force similar to that of WIMP, that is, weak nuclear interaction, have an abundance equal to that measured for dark matter. This "coincidence" is indicated by the term *WIMPs miracle*.

WIMPs have been the most fashionable particles for decades, but things have changed a bit in recent years. The reason is that the search for supersymmetry at the Large Hadron Collider (LHC) in Geneva has not produced positive results. The non-observation of supersymmetry could be due either to the fact that nature is not supersymmetric or to the fact that the energy to which it manifests itself is larger than that which was thought, and which can be scanned by the LHC, which in recent years has operated at an energy of 14 TeV.[2]

Now, if in the ordinary world of low energies we observe only usual, non-supersymmetric particles, this means that

---

[2] 1 TeV equals a thousand billion electronvolts.

there is a limit energy above which the world is supersymmetric and below which it is the ordinary one. This energy scale is the *breaking scale of supersymmetry*. Since supersymmetry solves the hierarchy problem, it is expected that this scale is not much greater than the electroweak scale, that is, a few hundred GeV. Unfortunately, experiments show that this is not the case. A higher-energy supersymmetry loses many of its attractive features such as that of solving the hierarchy problem.

7.3 Lost in Extra Dimensions

Other candidates for dark matter come from other areas of physics, such as scenarios of *extra dimensions*. In our daily life, we experience 4 dimensions: 3 spatial dimensions and a temporal one. Theoretical reasons have pushed physicists to add extra dimensions so small that they cannot be seen.

One of the fundamental objectives of modern physics is the search for a theory of everything that should bring all interactions together in a single framework. Although it is possible that this framework does not exist, there is still a sort of aesthetic sense that pushes people to seek such a unified description of the phenomena. This research was initiated and not concluded by Einstein, who was followed by several generations of physicists. In 1919, four years after the publication of general relativity, Theodor Kaluza proposed a theory of unification of electromagnetism and general relativity in a five-dimensional space, consisting of four spatial and one temporal dimensions. Kaluza sent the study to Einstein, who initially did not understand the role of the fourth spatial dimension, but then did his utmost to present it in 1918 at the Berlin Academy of Sciences, and to have it published in 1921. In 1926 Klein expanded the study by introducing quantum concepts, managing to explain the quantization of the electric charge based on the quantization of the moment in the extra dimension, and also calculated the order of magnitude of the extra dimension, which must have been about 10^{-22} m. After an initial interest, *Kaluza–Klein theory* was almost forgotten, until the seventies.

A descendant of the Kaluza–Klein theory, in a 10-dimensional space is the *superstring theory*, which is proposed as a possible theory of everything, in which particles are generated by the vibrations of tiny strings. Initially, string theory only included bosons, in 26 dimensions, and it was a theory of strong interactions. Subsequently, Ramond, Neveu, and Schwarz showed how to introduce fermions. The connection of that theory with supersymmetry led to superstring theory. In 1974 Scherk and Schwarz also pointed out that this model, rather than a theory of strong interactions, was a theory of quantum gravity and more generally a *theory of everything*. A problem with the theory is that it works in 10 dimensional spaces, and our world has at most 4 dimensions. Where are the dimensions we don't see?

As Klein had already explained, the extra dimensions are compact, closed on themselves, and extremely small. To explain the idea of the extra dimension, let's consider a long cylinder, or a rope. If we observe the cylinder from a great distance, we will perceive it as a one-dimensional thread. To see the extra dimension, you have to look at it closely.

A tightrope walker can only move in one dimension, back and forth, but an ant can also turn around the cable, that is, around the base of the circle (Fig. 7.3).

One can have a single extra dimension or more than one that can also have periodic properties. Reality in the world with extra dimensions can show very curious, absurd, and comic aspects. As the dimensions increase, the "abilities" of living being increase. A one-dimensional world can be populated by linear, filiform objects. If one of them meets another, in front of him, he can no longer move unless he convinces the other to move in the opposite direction. If we go to two dimensions, the first problem disappears. Our living being can surpass the other, and if we move to three dimensions it can even surpass it by jumping over it. The possibilities, the capacities, increase with the increasing number of dimensions. A being in four dimensions could "teleport" himself between distant places and have other abilities that would make him appear to us as a superhero. In our world the reflection of the right hand gives

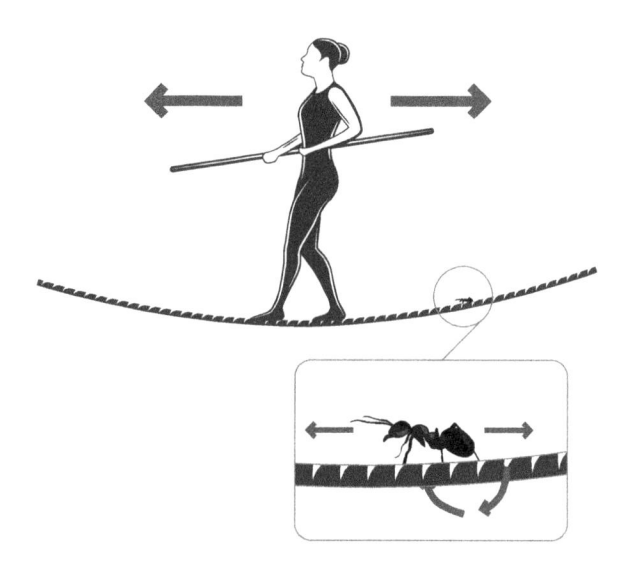

Fig. 7.3. Example of the concept of extra dimensions. For a tightrope walker, the wire has only one dimension, but for an ant it has two, because it can turn around it.

the left hand, while in a four-dimensional world the reflection is replaced by a rotation. So the glove of our right hand that visit the fourth dimension and comes back could be worn on the left hand. There are even tales that describe the journey into higher-dimensional worlds, such as *The Plattner Story and Others* by H. G. Wells, in which a teacher, upon returning from a trip to a parallel world, finds himself with the body parts reversed. Another example is *Flatland: A Romance of Many Dimensions* by E. Abbot, in which an inhabitant of a two-dimensional world comes into contact with one of a three-dimensional world.

Physics in a space also depends on the number of dimensions. In our world, an object can rotate around an axis, but, for example, in a four-dimensional world, it can rotate around two axes. The gravitational force decreases with the cube of the distance and not with the square, as in ours. These are just simple examples. We could go on and on. In short, the extra dimensions have effects on how we see reality, on physics, and on the mass and charge of particles.

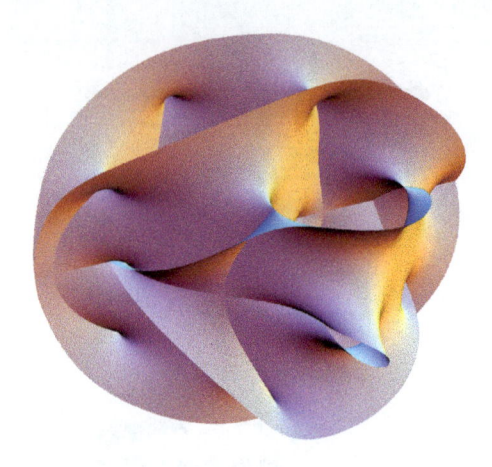

Fig. 7.4. Spaces of Calabi–Yau.
Source: Wikipedia.

Returning to the superstring theory, the extra dimensions are rolled up into complex figures called *Calabi–Yau spaces* (Fig. 7.4), six-dimensional spaces, which take their name from the Italian Eugenio Calabi and the Chinese Shing-Tung Yau, and are associated with each four-dimensional space-time point. Figure 7.4 gives an approximate idea of what this space is, approximate and distorted because we are viewing a six-dimensional space on a two-dimensional sheet of paper.

The number of possible compact manifolds of extra dimensions is huge — of the order of 10^{500} according to a 2004 estimation by Douglas, or even larger according to other authors. If we don't know how to compactify the extra dimensions, we can't make predictions for the real world on the basis of string theory.

So, since the number of ways of compactifying dimensions is enormous, string theory, which proposes itself as a theory of everything, actually becomes a theory of anything, because it does not explain our universe. String theory made many followers in the early stages. It made us hope that we were close to solving the age-old problem of the unification of gravity and quantum mechanics. Time has shown that things are not so trivial. There is little or no evidence that even string theory is a solid scientific theory. It provides supersymmetry, but the latter has not been observed at CERN. It reobtains Hawking's radiation, which, however, has never been observed, even if it is considered established physics. If the proton decays, this would be evidence in favor of supersymmetry, but, as discussed below, the proton does not decay, at least for 10^{34} years. Supersymmetry provides dark matter, but even this has never been observed. Other predictions of the theory are the existence of new long-range weak forces, also never observed. There was the hope that the theory could resolve the huge discrepancy between the observed and calculated value of the cosmological constant, a factor 10^{120}. Supersymmetry has been only able to reduce the discrepancy to values of the order of 10^{50}, but not able to solve the problem. However, people working in the field have not lost hope, and the search continues.

Going back to the extra dimensions, if we wanted to enter one, it would not be enough that we were small enough to reach the goal. Because of the quantum mechanical effects, we would need an energy inversely proportional to the radius of the extra dimension. Since this is very small, the energy needed would be enormous. So two particles with insufficient energy to penetrate into the extra dimension would only move on the surface of the cylinder, that is, there would only be one-dimensional motion. If we had two particles with enough energy to enter the extra dimension and collide, after the collision they could move to the extra dimension y, the base of the cylinder, in opposite directions. An observer who does not see that dimension would see that the particles collide and remain still. Consequently, it would be wrong to conclude that the principle of energy conservation does not apply. In reality, the two particles move in the extra dimension and the energy is conserved (Fig. 7.5).

Therefore, clues could be found to the existence of extra dimensions in particle accelerators by studying the apparent loss of energy due to motion in the extra dimension itself. Another interesting aspect is that for an observer who was not aware of the extra dimension, two particles that turn on the circumference y, the base circle, would be perceived as a single

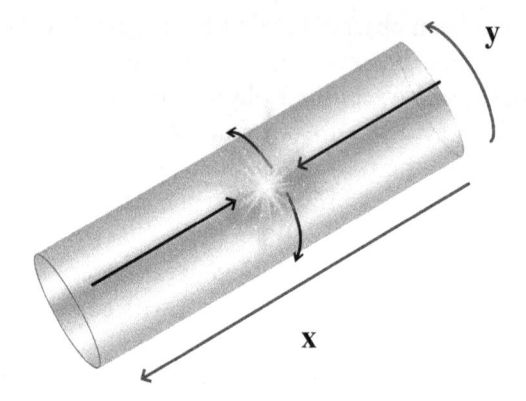

Fig. 7.5. In the image, the cylinder represents space. Usual dimensions, like those we experience in daily life, are represented by the straight dimension x, along the cylinder. The extra dimension, dubbed y, corresponds to a circle (compact cyclic dimension). In order for the extra dimension, y, to not be perceived, its dimension must be much smaller than the other dimension, x, and must be microscopic.

still particle, but of considerable mass. Actually, remembering the mass–energy relation ($E = mc^2$), the mass of the particles would be kinetic energy. So, according to the Kaluza–Klein theory, high-speed motion in extra dimensions generates what is the rest mass. In other words, mass can have two different interpretations: energy of a particle at rest, in accordance with special relativity, or energy due to motion in an invisible dimension. Another important point is that a particle would not appear to us as a single particle, but as a set of particles of different masses. The mass of each particle is related to the mass of the particle at rest in the extra dimensions.

So, any particle of the standard model that is in the extra dimensions could generate heavier particles, not yet discovered: these particles constitute "towers of particles"[3] of increasing mass. So, if we have an electron in motion in extra

---

[3] That is, the least massive particle, of mass m, is that of the standard model and the others have mass $m_n = \sqrt{(m^2 + n^2/R^2)}$, where for $n = 0$ we get the particle of the standard model and for $n = 1, 2 \ldots$, the other, most massive particles. As one can see the mass, m, depends on the dimension, R, of the extra dimensions.

dimensions, this will generate a *Kaluza–Klein (KK) tower* of heavier electrons.

Similarly, in superstring theory, the mass of elementary particles is due to their motion in invisible extra dimensions.

There are 5 different theories of superstrings, and they can be considered as 5 different aspects of a single theory called *M-theory*, in an 11-dimensional space. M-theory was introduced by Edward Witten in the so-called "second string revolution." The curious thing is that it is not known what the M indicates, because Witten never revealed it. Witten's solution to the problem of five theories instead of one is similar to the old Buddhist parable of the blind who were invited by a king to touch various parts of an elephant. Obviously, conflicting ideas of the elephant were made. Witten explained to confused physicists that each of the five theories is a manifestation of a single theory, the M-theory.

Another interesting point is that noted by Joseph Polchinski: string theory foresees and needs for its consistency the existence of larger objects such as *D-branes*. For example, a 2-brane is a membrane, a 0-brane is a particle, while a 1-brane is a string. The extension of string theory to branes complicated the problem of compactification even more, as it led to discover other ways to compactify the extra dimensions.

According to some researchers, our universe could be located on a 3-brane, having 3 + 1 dimensions, that shakes, like a sheet in the wind, in an environment of higher dimensions.

There are other peculiar models of extra dimensions.

In 1998 Arkani-Hamed, Dimopoulos, and Dvali (ADD) introduced the *theory of extra-large dimensions*. They showed that in a four-dimensional brane containing our universe incorporated in extra dimensions, they could solve the hierarchy problem and understand the extreme weakness of gravity compared to other fundamental interactions. In the model, the extra dimensions are compactified to form a circle with a tiny radius. The fundamental interactions of the standard model would be related to a four-dimensional space, and generated by open strings with their extremities on the brane (Fig. 7.6). This four-dimensional space is incorporated

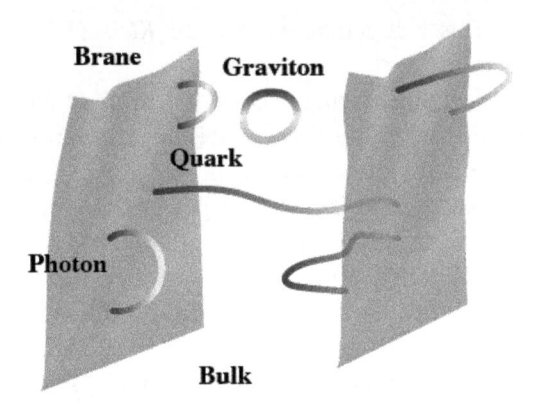

Fig. 7.6. Brane world. The open strings attached to the branes correspond to the material particles and the electroweak and strong interactions. Closed strings are the carriers of gravitational interaction that can move in the volume (bulk) between the branes.

into a larger space called *bulk* (the space between the branes) (Fig. 7.6). An analogy, to clarify the situation, is the two-dimensional surface of the Earth that is located within the three-dimensional space. In the analogy, the brane would be the Earth's surface.

Gravity carried by closed strings could move in outer space, and this would explain why it is so weak.

The assumption that gravity "lives" in a larger space-time leads to deviations from the Newtonian theory or general relativity, at small distances. This could be used as a test of the existence of extra dimensions, if it were possible to test gravity on very small scales.

A different model with extra dimensions that warp between two branes, was proposed in 1999 by Lisa Randall and Raman Sundrum. Such a model is known as *Randall–Sundrum theory* or *5-D[4] warped geometry theory*.

This theory assumes that the real world is an extradimensional universe described by a warped geometry, where the universe is constituted by a five-dimensional peculiar space, dubbed *anti-de Sitter space*. The standard model particles are

---

[4] 5-D stand for 5-dimensional.

localized on a (3 + 1)-dimensional brane. This theory has shown new ways of solving the hierarchy problem, but it has made people lose hope of finding a way to determine the right compactification. Remember, again, that there are at least 10^{500} different ways of compactifying the extra dimensions.

A characteristic of the extradimensional theories is that, by compactifying the extra dimensions, the fields that propagate in the bulk give rise to a series of oscillations called the *Kaluza–Klein tower of states*. In other words, particles that move in extra dimensions (4-branes) appear as heavy particles, a kind of copies of normal particles (Kaluza-Klein states). The particles we know represent the lowest level of the tower. In the scenario of *universal extra dimension* (UED) proposed in 2001 by Appelquist, Cheng, and Dobrescu, all the particles of the standard model can move in the extra dimensions, and once again each particle corresponds to its own Kaluza–Klein tower. The UED scenario as well as the ADD scenario would give observable effects in accelerators such as the LHC, such as the production of two particles of the standard model or more complicated signals. Experiments have been carried out at CERN, but they have failed to find such particles.

7.4 "Special Wanted"

The dark matter candidates constitute a real zoo. We have focused more on some specimens belonging to the WIMPs species because in the past years they received — and continue to receive — considerable attention. Such supersymmetric particles have merits, as mentioned earlier. Unfortunately, they have a big problem. After supersymmetry was not observed in the LHC experiments, doubts began to arise both on the existence of supersymmetry and, obviously, on its particles.

Even if supersymmetry and its particles do not exist, there are always a myriad of candidates for dark matter, such as *axions, sterile neutrinos, Wimpzillas, dark photon, fuzzy dark matter, Q-balls,* among others. There are also alternative scenarios to those of WIMPs and cold dark matter-like scenarios *such as that of self-interacting dark matter, asymmetric dark matter, fuzzy dark matter,* and so on.

The axion was proposed by Steven Weinberg and Frank Wilczek, and its name, chosen by Wilczek, comes from a well-known American detergent. This particle was introduced to solve a problem in nuclear physics. It is a particle with a mass smaller than one billionth of that of the electron, between 10^{-6}–10^{-3} eV. It has no spin, and it interacts very weakly with ordinary matter. For this reason, it is also referred to as "the invisible axion."

An important feature for the detection of axions is the fact that they can turn into photons in the presence of intense magnetic fields. Axions were produced in the primordial universe with a thermal mechanism such as that of WIMPs or with a completely different non-thermal mechanism.

Sterile neutrinos interact with other particles only through gravity. They were introduced to find a solution to the neutrino mass problem, as described in Chapter 6. The mass of right-handed (see Appendix D) or sterile neutrinos is unknown and could vary from less than 1 eV to 10^{15} GeV, and they could be a solution to the problem of missing mass.

Another possible form of dark matter are the *Wimpzillas*, whose name was obtained by adding the terminal part of the word Godzilla to WIMP. The name is obviously due to their huge mass: 10^{12}–10^{19} GeV. They were introduced to explain the existence of cosmic rays with very high energies.

Q-balls, where the Q stands for electric charge, are localized spherical field configurations.

The *hidden or dark sector* is a kind of collections of quantum fields, and their particles have not yet been observed. The interactions between the dark sector particles and those of the standard model are weak, mediated through gravity or through the so-called *dark photon*, axions, or sterile neutrino. Even the Higgs boson could interact with the dark sector particles and act as a communication portal, called the *Higgs portal*, between the standard model and the dark sector.

As for the *self-interacting dark matter* scenario, it is linked to problems of the WIMPs model. On small scales (1 kpc), WIMPs have problems reproducing the structure of galaxies. There are a number of problems — for example, the so-called *cusp–core problem*. The distribution of dark matter, especially in dwarf galaxies, has a flat density in the center, while simulations using WIMPs produce very steep density profiles, called *cuspy profiles*. To this problem, others are added: the *missing satellites problem*, that is, in our galaxy and Andromeda, fewer satellite galaxies are observed than those predicted by the WIMPs model.

The *too-big-to-fail problem* is related to the discrepancy between the central density of galaxies predicted by the WIMPs model and observations. Another problem is related to the observation of the distribution of satellite galaxies on a plane structure, for example in Andromeda, which is not compatible with the WIMPs model. Possible solutions to these problems are related to the role of the interaction between dark and visible matter which would change the distribution of dark matter on small scales. Another possibility is to assume that dark matter is *self-interacting* or *fuzzy dark matter*, this last constituted by extremely light scalar particles (bosons with spin equal to zero) having masses of the order of 10^{-22} eV.

Another scenario proposed is that of the *asymmetric dark matter*. The basic idea is that since the amount of ordinary matter and that of dark matter are similar (dark matter is only a factor of 5 greater than ordinary matter), the density of dark matter today would have originated because of a process similar to that of antimatter–matter asymmetry. So dark matter could have been created in some process involving an *asymmetry between dark matter* and *anti-dark matter*.

The number of dark matter candidates constitutes a real zoo. The search has been mainly centered on WIMPs, but also on axions and sterile neutrinos. As we will see in the next chapter, to date, no one in this zoo of candidates has shown a clear sign of its existence.

Chapter 8

DETECTION OF DARK MATTER

Science progresses best when observations force us to alter our preconceptions.

— Vera Rubin

8.1 How to Find Dark Matter

In 2012, Katherine Freese and Christopher Savage published an article about dark matter collisions with the human body, showing that of the billions of dark matter particles that pass through us every second, a WIMP can hit one of the quarks that make us up with the exchange of a Higgs boson, and in a year, about ten WIMPs interact with our atoms. In other words, although the interactions of dark matter particles with ordinary matter are so weak, they are not null.

Obviously, these particles pass through and can interact with all objects. Thus, to look for dark matter, we could build an experiment that tries to verify if WIMPs have interacted with nuclei. This method of researching dark matter, already proposed in the 1980s, is called *direct detection*. The probability of a dark matter particle to interact with a detector depends not only on its mass, but also on its density. Dark matter is found everywhere in space. Its density increases, going towards the center of our galaxy. Around the Earth, about 8 kpc from the galactic center, most of the mass is baryonic, which means ordinary matter. In any case, a coffee cup contains some dark matter particles that do not stand still but move at high speeds, of the order of 300 km/s.

In addition to this method, there are two other dark matter detection methods, the first based on the decay of dark matter particles into different possible particles, called *indirect detection*, and the other based on the possibility that dark matter is generated and observed indirectly in *particle accelerators*.

8.2 Close Encounters

In the Max Planck Institute in Munich, Andrzej Drukier and Leo Stodolsky studied techniques that allowed to reveal the elastic scattering of neutrinos from nuclei. In 1984 they published an article describing how a *superconducting material*[1] could be used for the purpose. The following year, Mark Goodman and Ed Witten proposed to use the same technology for the detection of dark matter, discussing scattering events independent of the spin of the nucleon and others dependent on it. The *spin-independent scattering* is particularly interesting because the cross section, that is, the probability of interaction, increases with the square of the mass number, that is, the sum of the number of protons and neutrons in the nucleus. Theoretical calculations showed that for an interaction of this type, in a day, and for each kilogram of detector, a number of events up to a thousand could be generated.

In summary, direct detection is based on the idea that a dark matter particle interacts with the nucleus of a certain material, producing a recoil of the nucleus that deposits energy in the material which is then measured in different ways (Figs. 8.1, 8.2).

A first possibility is the measurement of the vibrations of a crystal lattice, *phonons*,[2] a quantum of oscillation, similar to the "photon," which is a quantum of light. Phonons are not mysterious entities. A phenomenon similar to that of phonons occurs when one beats a bar of a given material (preferably a

---

[1] A type of material that is able to conduct electricity with no resistance.
[2] Phonons are oscillations of the internal structure of a material.

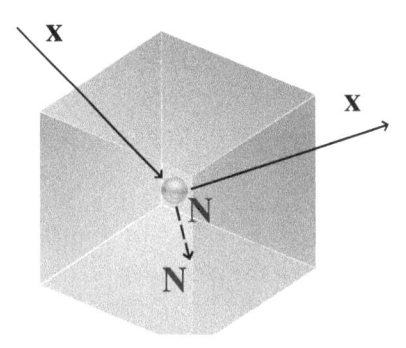

Fig. 8.1. Direct detection. A dark matter particle, X, hits a nucleus, N, which recoils and releases energy.

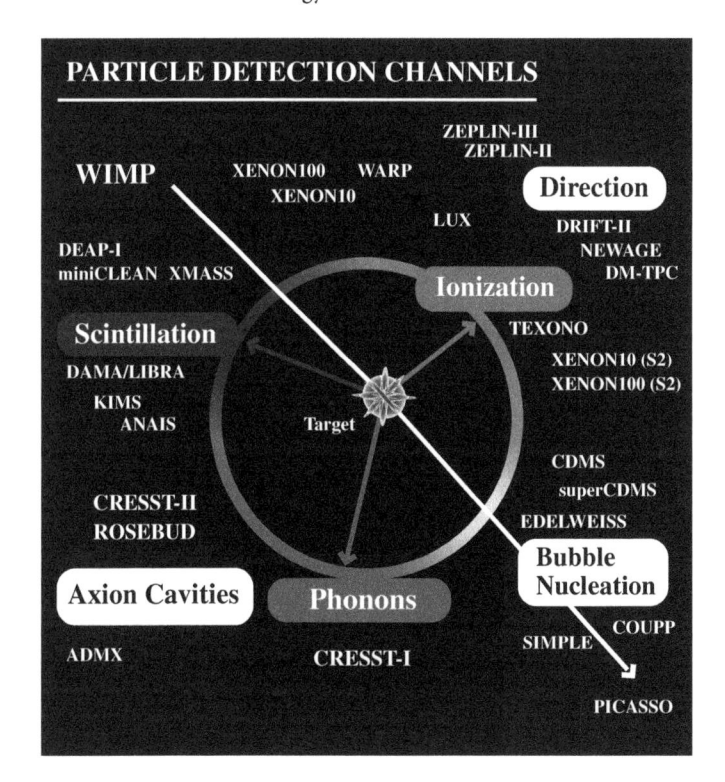

Fig. 8.2. Experiments and methods of direct detection.

metal such as iron) with some object. Touching it with one hand, you can feel oscillations, which are a sort of amplified version of the phonons. The second detection possibility uses the *ionization detection*, namely the gaining or losing of electrons making an atom charged. The third uses *scintillation*, that is, the generation of light pulses emitted after a collision. There are dozens of direct detection experiments, as shown in Fig. 8.2, that measure one or more of the indicated effects, dubbed *detection channels*.

Historically, the first test of the theory was carried out in 1986 in the Homestake mine in South Dakota, where experiments were started to get an idea of the scattering cross-section of dark matter with the nuclei using a germanium ionization detector.

One of the major problems in these types of experiments is the other sources of interaction with the detector cores, such as natural radioactivity from the environment near the detector (α, β, γ decay, neutrons) and cosmic rays. The latter are relativistic, high-energy particles and atomic nuclei coming from the cosmos, hitting the Earth constantly. Consequently, they also hit the detector cores.

Because of this sort of background noise, if we tried to carry out one of those experiments on the surface of the Earth, events of collisions of dark matter with the nuclei would be submerged. It is necessary to try to block or decrease the bombardment of the particles coming from the cosmos. To do this, the detector must be shielded with kilometers of rocks, as in the Gran Sasso National Laboratory. To try to control the effects of the background (background noise) that introduces "false alarms," one can change the detection method and use the one proposed in 1986 by Andrzej Drukier, Katherine Freese, and David Spergel. The Earth rotates around the Sun with a variable speed and direction, and the Sun moves around the Milky Way in the direction of the Cygnus constellation. Consequently, an observer on Earth "sees" a wind of particles coming from that direction. A detector should therefore observe a greater flow as the alignment of the speed of the Earth with solar motion increases (Fig. 8.3). The maximum flow is therefore expected in June. This is what is observed in one of the experiments that study dark matter: DAMA/NaI, second generation of the DAMA/LIBRA experiment.

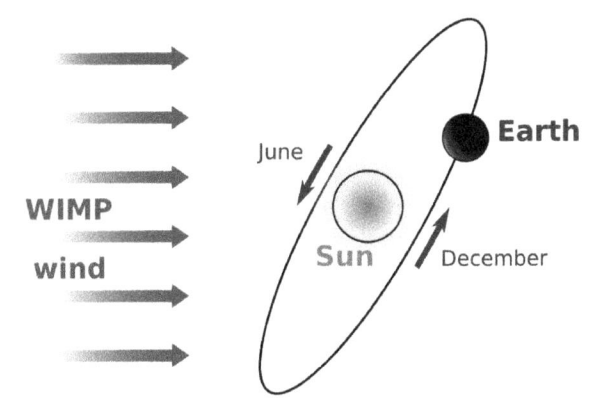

Fig. 8.3. Solar system motion towards the Cygnus constellation against the WIMPs wind. Credit: K. Freese, M. Lisanti, and C. Savage, "Annual Modulation of Dark Matter: A Review."

The experiment is located underground, in the Gran Sasso National Laboratory, and the detection method or channel is the *scintillation of the sodium iodide crystals*. Since 1998, the start date of the DAMA experiment data collection, until today, the seasonal variation of the signal has reached a very high level of significance (more than 9 σ). Physicists use the σ to determine how much a result is due to chance, or not. For example, 3 σ, corresponds to a probability of 1 in 740, and 5 σ corresponds to a probability of 1 in 3.5 million. As you can see, the higher the σ, the lower the probability that the result is due to chance. 9 σ corresponds to about 8.9×10^{18}, that is, the probability that the result is due to chance is almost zero. Experts are very skeptical of the results of the DAMA/LIBRA collaboration, albeit of very high significance. The skepticism stems from the fact that other experiments such as XENON, as we will see, have reached far greater sensitivities than that of DAMA/LIBRA, but have seen no signal of dark matter. The collaboration carried out detailed studies to exclude that the signal came from other sources with a negative result. Independent studies assume that the seasonal variation may be due to the seasonal variations of the muon or neutron background.

Ultimately, it is not clear whether the signal measured by DAMA/LIBRA is really related to dark matter or not. If it were, the discrepancy with the other experiments could indicate our poor knowledge of how dark matter interacts with ordinary matter. In an attempt to answer the questions raised by the DAMA/LIBRA collaboration, the DMICE experiment is under construction in the Southern Hemisphere, using the same types of crystals and the method of seasonal variations. In this way, it will be possible to determine if the signal is linked to the instrumentation or to environmental effects.

During the period in which DAMA was developed, new techniques able to discriminate between the dark matter signals from those of different types of backgrounds were proposed. In 1985 Cabrera, Krauss, and Wilczek proposed an experiment that by measuring the temperature variation due to the interaction of neutrinos with an experimental device could reveal these particles. Cabrera and Sadoulet extended this idea to the search for dark matter. The experiment that came out of it was called CDMS[3] (Cryogenic Dark Matter Search) and based on germanium crystals cooled to temperatures close to absolute zero (−273.16°C). This experiment is part of a group of experiments called *cryogenic experiments*, such as COGENT, CRESST, and EDELWEISS.

The energy deposited in the crystal after the interaction is measured using one or two methods. COGENT operates in the *ionization channel*. CDMS measures *ionization and phonons*. As already mentioned, phonons are oscillations of a crystal lattice of very low energy. This energy can be measured by bringing the crystal to a few thousandths of a kelvin and using a superconductor on the surface. Aside from phonons, and ionization, scintillation signals are also measured. By combining these channels, it is possible to discriminate between signals produced by dark matter and those of ordinary matter.

---

[3] CDMS has several "heirs": CDMSI, II, SuperCDMS, and SuperCDMS SNOLAB, the latter still under construction.

Some of the cryogenic experiments, such as COGENT, CDMS, and CRESST, which measure scintillation and released heat, had initially obtained more events than expected from background noise. Subsequent checks disproved the first results.

The probability of detection of dark matter is basically dominated by the product of the quantity of material used, called *target*, by the duration of the experiment, called *exposure*, and by the ability to reduce the background.

The experiments we discussed, such as CDMS, CRESST, and COGENT, used a quantity of material of the order of a kilogram. To increase the probability of interaction, hundreds or thousands of kilograms must be reached. Doing so with experiments based on crystals is expensive, and technology has switched to the use of noble liquids or gases.

8.3 Noble Gases

The technique based on the use of noble gases in the liquid state was introduced in the late 1990s by Pio Picchi, Hanguo Wang, and David Cline.

Noble gases are monoatomic, non-reactive, odorless, and colorless. The name comes precisely from the fact that they are not chemically reactive, and therefore do not tend to mix, like the nobles, with ordinary people. The most common in the atmosphere is *argon*, 0.932%. Atmospheric argon contains argon-40, with 40 protons, and traces of argon-39 which is radioactive and therefore introduces disturbing signals if it is used for the search for dark matter. If you want to use this gas for dark matter research, you need to restrict yourself to argon-40. This is the first step taken by the DARKSIDE collaboration using argon as a detector. It is extracted from the Colorado subsoil, then taken to Fermilab, where it is purified, and finally sent to the Gran Sasso National Laboratory. The DARKSIDE10 prototype was tested in 2012 and DARKSIDE50 started in 2013. DARKSIDE-20k, containing 20 tons of argon, is under construction and is one of the promises in the search for dark matter in the form of WIMPs.

One of the noble gases that has allowed a significant increase in the sensitivity of the detectors is *xenon* (from the Greek "foreign"). It was spotted in 1898 by William Ramsay and Morris Travers in England. It is a gas present in traces in the atmosphere, has 54 protons, and is used to make lamps and luminous devices, anesthetics, as well as in magnetic resonance, and in ion-propulsion engines. Experiments that proved to be particularly effective are LUX and XENON: XENON10, which used 15 kg of liquid xenon, XENON100, 165 kg of liquid xenon, XEXON1T, 3.2 tons of liquid xenon. The channels of detection are scintillation and ionization. When a dark matter particle interacts with a xenon nucleus, a flash of light is generated. It is measured by instruments called *photomultiplier tubes*. At the same time, electrons are ripped from the nucleus of xenon and move towards the top of the detector guided by a magnetic field. Above the liquid xenon there is xenon in the gas phase, in which the electric field is higher than in the liquid phase and the electrons are accelerated until the production of a further scintillation event. In this way, the detector allows a reconstruction of the three-dimensional position of the interaction of the particle. The correlations of the scintillation signals allow to establish whether there are events of interaction between dark matter and the nuclei of xenon, or between usual particles and the nucleus or electrons of xenon. The reduction of the background has grown from XENON10 to XENON1T, thanks to the choice of particular materials and to the shielding of the detector by a large water tank. Thanks to these measures, XENON1T could even reveal dark matter if there only was an event in a year per ton of xenon. For computers, Moore's law scales the progress of computing power every 18–24 months. Dark matter detectors follow a similar law: sensitivity increases by a factor of 10 in about two years, as shown in Fig. 8.4.

In the latest XENON1T measurements, the cross section (interaction intensity) of WIMPs with mass around 30 GeV is around 10^{-47} cm^2. XENONnT and DARWIN are the next

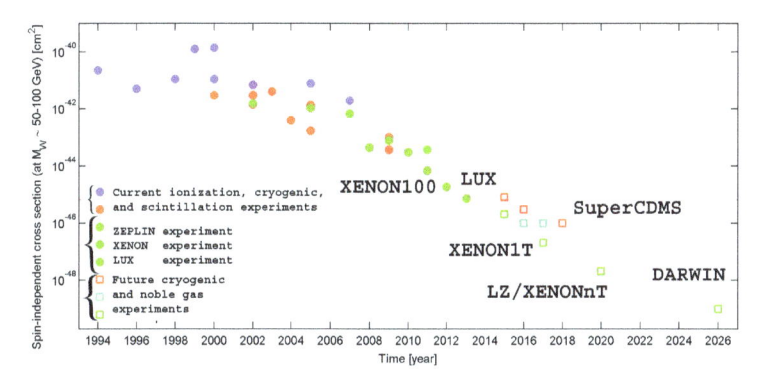

Fig. 8.4. Increase in the sensitivity of the measurements over time. Some current experiments and others to be carried out are indicated. Credit: Modification of work by Laura Baudis ("WIMP Dark Matter Direct-Detection Searches in Noble Gases").

xenon detectors that will push sensitivity even further. Despite the efforts made and the great sensitivities achieved, it has not yet been possible to reveal any dark matter particle. The absence of detection is, however, important because it gives indications on what dark matter is not. The results of the above experiments are usually represented in a plane, with the horizontal axis giving the mass of the dark matter particles and the vertical axis the intensity of interaction, as shown in Fig. 8.5.

The curves represented in Fig. 8.5, called *exclusion curves*, refer to different experiments and substantially exclude the upper area of the plane. Dark matter particles can only be found below them. In this graph, the *closed curves* indicate dark matter observation "claims," as in the case of the DAMA/LIBRA experiment. Several experiments with increasing sensitivity from top to bottom are shown. Some, like LZ, are under construction. Experiments continue to increase sensitivity, but this improvement cannot go on forever. For experiments a few hundred times larger than XENON1T, the scattering neutrino-nucleus, represented by the orange line, will begin to create problems. It will then be necessary to find a method to distinguish neutrinos from dark matter signals. Added to this, the possible lack of detection of dark

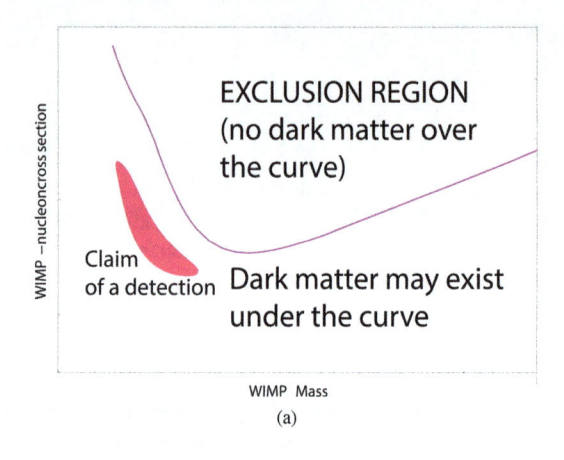

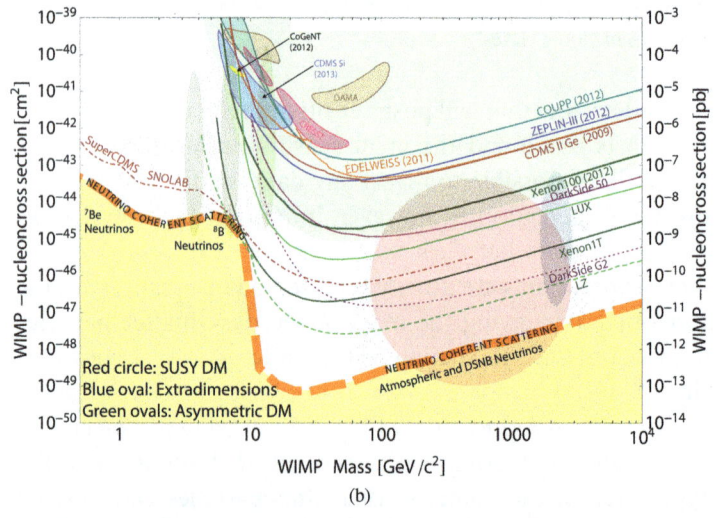

Fig. 8.5. (a) Sensitivity of generic experiments. The open line is an *exclusion curve*, thus named because it excludes the plane region above itself. The closed curves refer to experiments that claimed to have revealed dark matter, for example DAMA. (b) Sensitivity of planned and running direct detection experiments. The full curves are experiments that operated or are operating. Dashed lines are future experiments. The closed regions represent the allowed regions by DAMA (brown), CREST (pink), CDMS (blue), and COGENT (yellow) signals. The red circle is the region favored by SUSY models. The blue oval (between Xenon100 and LZ) represents the dark matter coming from extra dimensions, and the green ovals at the left represents the asymmetric dark matter. The line "neutrino coherent scattering" represents the (irreducible) neutrino background. Credit: T. Saab, E. Figueroa-Feliciano, and SuperCDMS Collaboration, "Dark Matter Limit Plotter," https://supercdms.slac.stanford.edu/dark-matter-limit-plotter.

matter in direct detection experiments and in the accelerators would lead to the abandonment of the WIMPs paradigm. At this point the remaining hope would be to consider that dark matter is made up of other particles such as axions. Recently, the collaboration XENON1T observed an excess of electron recoil, which has been interpreted as solar axions coupled to electrons.

One of the ways for axion detection is their transformation into photons in the presence of strong magnetic fields, known as the *Primakoff effect*. CERN runs an experiment named CAST (the CERN Solar Axis Telescope), looking for axions coming from the Sun. Another way of detecting axions is by using microwave resonant cavities inserted in superconducting magnets. This method is used in ADMX (Axion Dark Matter Experiment) located in the University of Washington, Seattle.

Another way to detect axions is to study the rotation of *polarized light*[4] passing through intense magnetic fields. In 2006, the PAVLAS collaboration, based on this method, claimed to have observed such a rotation. If true, this could indicate the existence of axions having a mass of about $1-1.5 \times 10^{-3}$ eV. Unfortunately, the result has been disproved by subsequent experiments. There are astrophysical phenomena in which axions could leave their "signature." In stars with intense magnetic fields, such as magnetars (magnetic stars), the conversion of photons into axions could be much more efficient than in the laboratory.

8.4 Indirect Detection: Dark Matter and Photons

Aside from the direct detection method, a second possibility is *indirect detection*. The foundations of this second method were published in two papers in 1978 by Gunn and collaborators and by Floyd Stecker. The two groups focused on heavy leptons (heavy neutrinos), showing how from the annihilation

---

[4] Light originates from oscillations of the electric and magnetic field. When the vector representing the electric field oscillates only in one plane, the light is polarized.

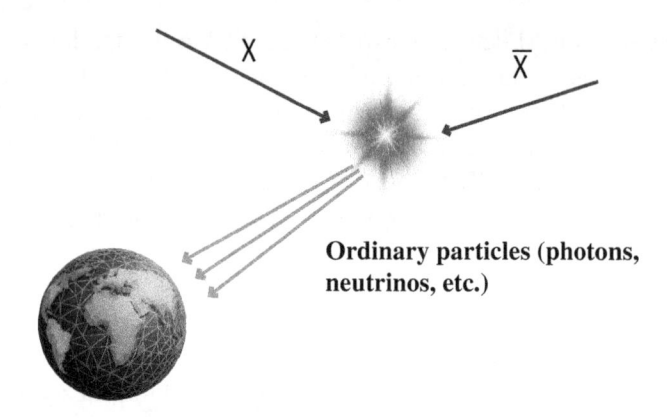

Fig. 8.6 Annihilation of dark matter particles, χ, that can be detected on Earth.

of pairs of these particles could originate a flow of gamma rays. The study was extended to dark matter candidates of various types, such as WIMPs. These particles annihilate, producing energies in the range of 1–1000 GeV (1–1000 times the mass of the proton) in the form of very energetic photons, that then interact with the nuclei present in the atmosphere, giving rise to cascades of particles (Fig. 8.6). To observe such gamma rays, there are two possibilities.

The first possibility is to use satellites that are above the atmosphere, while the second is to use terrestrial telescopes relying on a particular effect, which we will discuss. A significantly important satellite in this research is the *FERMI satellite*, conceived for the study of electromagnetic radiation in the 8 keV–300 GeV range. It was launched in 2008 and consists of a detector for gamma-ray bursts and a *large-area telescope* (LAT) that is sensitive to gamma radiation between 20 MeV and 300 GeV. In this second detector, the gamma photons are converted into electron–positron pairs, whose deposited energy is measured and the direction of origin of the photons is reconstructed. The study of dark matter starting from gamma rays has the advantage that these are not deflected by magnetic fields, as instead happens with charged particles, which allows to trace the source.

Terrestrial telescopes operate in a completely different way. When high-energy photons interact with the nuclei of the atmosphere, swarms of particles are generated, such as electrons and positrons, which move at a speed larger than that of light in the air, producing a blue radiation focused in a cone, dubbed *Cherenkov radiation*. This light is collected by telescopes with large mirrors. For example, the HESS telescope is made up of four telescopes and the surface of each mirror is 10^7 m^2. Mirrors are made up of pixel cameras that allow rather sharp cuts on the images. The *stereoscopic technique* (i.e., multiple telescopes observe the same event from different angles) traces the energy and the exact geometry of the event and improves the angular resolution. This technique combines the images, thereby allowing to determine the trajectory of the swarm of particles, and also the energy and the direction of the photons' arrival.

Energy is calculated from the intensity of the image; the direction, from the orientation of the image. The type of particle is obtained from the shape of the image. Nowadays, several such telescopes are operating: the already mentioned HESS (High Energy Stereoscopic System) in Namibia, VERITAs (Very Energetic Radiation Imaging Telescope Array System) in Arizona, MAGIC (Major Atmospheric Gamma Imaging Cherenkov Telescope) in the Canary Islands, Whipple in Arizona, CANGAROO-III in southern Australia, and HAWC (High-Altitude Water Cherenkov Observatory) in Mexico.

The future of the detection of dark matter that annihilates in photons is the CTA (Cherenkov Telescope Array), consisting of a few hundred telescopes of three different sizes, 4 m, 12 m, and 23 m, to reveal photons of different energies. This project is expected to be completed around 2025.

Not all regions are created equal for the purpose of detection of dark matter. The flow of gamma photons due to the decay of WIMPs mainly depends on the square of the density of the WIMPs and on the distance. This means that although structures such as clusters contain huge quantities of dark matter, the flow is diminished by their great distance. Conversely, dark matter clusters in our galaxy, commonly called *substructures*, are much closer than clusters, although weaker. To choose the regions of the sky to be studied, therefore, these effects must be taken into account. It should also be remembered that gamma rays can also be produced by astrophysical sources such as pulsars, which complicate research by acting as background noise generators. In the case of our galaxy, we know that most dark matter is concentrated in its center and therefore it is one of the most important regions to study. Studies of this region by several telescopes have highlighted an excess of gamma emission consistent with the annihilation of particles with mass of the order of 50 GeV. In the same region, however, there are many astrophysical sources whose gamma emission can be mixed up for those of the annihilation of WIMPs. Another problem is that simulations and observations have failed to establish what the real density of dark matter is in the central area of the galaxy. The distribution of dark matter in the galaxy is not uniform: there are some sort of lumps or substructures of varying sizes, whose position is unknown. Among the substructures, the obvious ones are the *dwarf galaxies*, such as the Magellanic Clouds that orbit our galaxy. Together with the galactic center, the dwarf galaxies have been studied by the FERMI satellite, which has set limits on the annihilation of WIMPs with mass less than 100 GeV, while terrestrial telescopes have studied the annihilation of more massive dark matter particles.

8.5 Dark Matter and Antimatter

Another method to research dark matter was proposed by Silk and Srednicki on the basis of the study of antimatter (positrons and antiprotons) in cosmic rays. As we have seen, antimatter is ordinary matter with some different properties from those of ordinary matter. For example, the antimatter particle corresponding to the electron, that is, the positron, has the same mass as the electron, but opposite charge. Antimatter is therefore not as exotic as seen in science fiction movies. It is very rare in our universe, but small quantities exist even in the human body.

Bananas contain potassium-40, in small quantities, which decays and emits a positron, every 75 minutes. When we eat a banana, we eat very small quantities of antimatter, so small that it does not cause problems to our health. Other antimatter constantly arrives on Earth in the form of cosmic rays and is produced even above thunderstorms.

What is the link between antimatter and dark matter? As seen, antimatter is contained in cosmic rays, with a ratio between positrons and electrons decreasing with increasing energy. In the annihilation of dark matter, however, the relationship between positrons and electrons grows with energy. So to look for signs of the existence of dark matter, one could see if in cosmic rays the relationship between positrons and electrons grows with energy. To this end, a group of Italian, Swedish, and German researchers launched a space research program in collaboration with Russian scientists. A set of Italian institutes prepared sophisticated detectors that, when assembled, formed the PAMELA (Payload for Antimatter Matter Exploration and Light-nuclei Astrophysics) instrument, launched on June 15, 2006, by a Soyuz. A couple of years later, in 2008, the growth between the relationship of positrons and electrons with energy was observed and published in a 2009 issue of *Nature*. This result was confirmed a few years later from the FERMI satellite. However, there were some problems. To explain the data, it was necessary that the annihilation rate of WIMPs be 100 to 1000 times larger than that produced by the decay of WIMPs, with the further consequence that an associated antiproton flow should have been observed, which however was not. The problem of annihilation rate could be solved by referring to local inhomogeneities in dark matter that would have produced an amplification of the annihilation rate. However, it has been shown that the signal discovered by PAMELA could also be explained in terms of astrophysical sources such as pulsars. This is the major problem of indirect detection: the difficulty in distinguishing between astrophysical and dark matter signals. Other instruments after PAMELA, such as FERMI, AMS-02, installed in 2011 on the International Space Station, CALET (Calorimetric Electron Telescope), and DAMPE (Dark Matter Explorer), launched in

2015, continued and continue the search for dark matter with results, alas, so far not positive.

8.6 Dark Matter and Neutrinos

To overcome the problem of distinguishing astrophysical signals from those of dark matter, ultra-energetic neutrinos can be used. Dark matter permeates the entire universe and is subject to the force of gravity. Objects generating an intense gravitational field such as neutron stars, and to a lesser extent our Sun, can attract dark matter particles that remain trapped by the gravitational field and colliding with the solar nuclei lose energy and settle in the central part of the star.

Neutrinos generated by the annihilation of WIMPs can leave the Sun and could be revealed on Earth. This method was proposed in 1985 by Krauss and collaborators, and to put it into practice several experiments were built: BDUNT (Baikal Deep Underwater Neutrino Telescope) built inside the Russian lake Baikal, ANTARES (Astronomy with a Neutrino Telescope and Abyss environmental Research project) in the south of France, KM3NET under construction 80 km southeast of Capo Passero, south east coast of Sicily, at a depth of 3500 m. Other experiments were built in the ice of the South Pole: AMANDA and IceCube, made in a cubic kilometer of ice between 1500 and 2500 m of depth (Fig. 8.7). The basic principle is common to all experiments. When ultra-energetic neutrinos from the Sun give rise to electrons or muons in water or ice, which move at speeds larger than that of light, they produce a flash of Cherenkov light. This radiation is detected by optical modules containing photomultipliers tied to strings dropped into water or ice.

Unfortunately, even these experiments have so far not yielded any results.

8.7 A Recipe for Cooking Dark Matter

There is a third way to reveal dark matter: that of producing it in an accelerator like the LHC, from collisions at very high

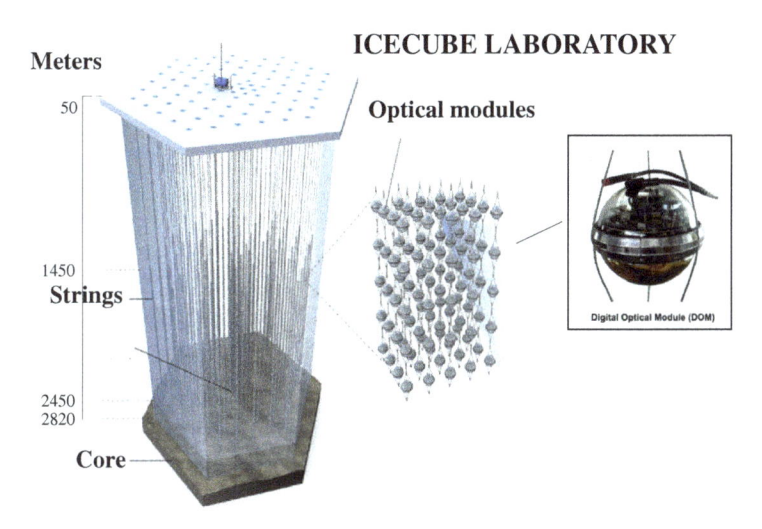

Fig. 8.7. Ice cube scheme. It consists of 86 strings in the ice. Optical modules, consisting of photomultipliers that observe Cherenkov light, are attached to each string. Credit: Adapted from the IceCube website (https://icecube.wisc.edu/science/icecube/detector).

energies between ordinary particles. The LHC is the largest and most powerful hadron accelerator in the world with an energy of around 14 TeV. It is located inside a 27 km tunnel 100 m deep below Geneva, between Swiss and French territory. Although dark matter interacts very weakly with ordinary matter, there is the possibility of generating dark matter particles from the collision of ordinary particles, for example two protons. An important part of the LHC's program is to reveal non-ordinary particles. We have already talked about supersymmetric particles, and as we know, WIMPs are supersymmetric particles. Their discovery would have the same or greater relevance than the discovery of the Higgs boson. Making two protons collide at high energies would generate a particle of dark matter and its antiparticle. The presence of the dark matter particle could be inferred from the fact that in the final events there would be problems with the conservation of energy and moment. This because the dark matter particle would interact so weakly with ordinary matter that it would be created and would fly away unnoticed, but carrying its mass-energy with it. This produces a violation of the principle of energy conservation.

In more detail, interesting events are given by reactions such as the following:

$$pp \to \chi\bar{\chi} + x$$

where two protons colliding generate a particle of dark matter, χ, and its antiparticle plus something else (indicated by x), which can be a photon, the W and Z bosons in leptonic decay, or a hadronic jet, which is a narrow cone of particles generated by the formation of hadrons from quarks and free gluons. In the CMS and ATLAS experiments of the LHC, signs of the existence of new particles relating to different supersymmetric and extradimensional models were sought. The results were all in agreement with those of the standard model.

For precision's sake, in 2015, the LHC observed a signal interpretable as the decay of a hypothetical 750 GeV particle in two photons. It was later shown that it was a statistical fluctuation.

8.8 Experimental Evidence of the Existence of Dark Matter

From the previous discussion, we know that we have remarkable paraphernalia for the detection of dark matter. In recent decades, several claims have been made about its existence, all inexorably disproved. We have already seen a couple of them. The evidence from PAMELA's data was shown to be compatible with astrophysical signals. The detection of a hypothetical 750 GeV particle was shown to be a statistical fluctuation. Besides these examples, there are several others that we want to remember here. In 1994–95, the HEAT balloon observed an excess of positrons in the 7–30 GeV range, later confirmed by PAMELA and extended up to 200 Gev by FERMI. In 1997, the EGRET space telescope observed an excess of gamma rays above 1 GeV compared to the expectations of galactic cosmic ray models, consistent with a WIMP of mass 50–100 GeV. In 2003, the INTEGRAL space telescope observed a bright emission at 511 keV from the bulge of our galaxy. WMAP and then PLANCK observed a *"haze"* along the plane of our galaxy, interpreted as produced by a 100 GeV WIMP. The FERMI satellite had observed an emission orthogonal to the galactic plane in the shape of two symmetrical balloons called *FERMI bubbles*. In 2014, the same satellite observed an excess of gamma emission in a part of the region. A 2015 publication showed that there was actually no sign of dark matter.

All these results do not give any certainty about the existence of dark matter. As seen, the dark matter detection claims are not convincing, sometimes because of the limitations of the search device, sometimes because of the difficulty of distinguishing between astrophysical effects that can mimic the effects of dark matter (e.g., the case of PAMELA), sometimes because in the experiment only statistical fluctuations are observed (e.g., the case of the 750 GeV particle).

An incontrovertible proof of dark matter existence would be the observation of a *gamma ray line*, since no astrophysical processes are known to be able to produce a *line* at energies greater than 1 GeV. In 2013, FERMI observed a 133 GeV line, but subsequent studies showed a low statistical significance of the signal. In 2014, the XMM-Newton and Chandra satellites observed a 3.5 keV line, interpreted as possible decay of sterile neutrinos or axions. Unfortunately, the signal could be due to astrophysical processes, or it could simply be a line produced by plasma.

Although there are no positive results from the experiments, they have contributed to excluding dark matter models and to narrowing the field of possibilities.

8.9 The Future of Dark Matter Research

In conclusion, we have several types of experiments that theoretically allow the detection of dark matter particles. A question that can be asked is, is it enough for only one of them to give a positive signal for us to be sure we have found dark matter? The answer is not so trivial.

For example, if new particles could be produced in the LHC, it would be necessary to ensure that they have all the characteristics of dark matter. For example, they should be stable. Checking this in the LHC is not trivial, since we could only determine the stability over very short times corresponding to the time that the particles take to be generated and revealed. Another problem is that we could not know if the new particles are abundant enough to constitute all the dark matter in the universe. In other words, the LHC experiments alone would not tell us with certainty that new particles make up dark matter.

Similarly, if one of the direct detection experiments observed a signal higher than the background noise, we could not know for sure if this signal was actually due to dark matter or some unaccounted background noise. We could not even know if the particles that originated the signal are in the right amount to constitute all dark matter. Finally, the signals in indirect detection experiments interpretable as dark matter could be due to astrophysical sources, and this is what happened with several of the results obtained in the past, for example with PAMELA, to remember just one.

In conclusion, to reveal dark matter we would have to be very lucky and it would be necessary that more than one of the quoted techniques give positive results, so that we could cross the "identikit" given by one technique with that of the other and hope that they are compatible. The search for dark matter has similarities with the story of the blind and the elephant, already told. Like the blind men, to try to reveal it and understand its nature, it is necessary to gather information from different points of view (with different tools). Unlike the blind people who ended up arguing about the characteristics of the elephant, the information received with the different methods must be combined in the right way to solve the puzzle.

However, with our current knowledge, we are left with a doubt: what if dark matter did not exist? We will discuss this in section 9.6

Chapter 9

DARK ENERGY

Truth is stranger than fiction, but it is because Fiction is obliged to stick to possibilities; Truth isn't.

— Mark Twain

9.1 A New Surprise of Cosmology

Contrary to the conception of a static universe that led Einstein to modify the equations of general relativity, Friedman and Lemaître showed that Einstein's equation predicted an expanding universe. However, it was not until 1929, after the publication of the Hubble results, that the idea that the universe was expanding was finally accepted. The famous *Hubble's law* showed that Earth and our galaxy were not in a privileged position in the universe. This had been a great discovery that had changed the paradigm of the static nature of the universe that everyone was sure of. However, this was not the last big surprise that cosmology reserved to us. Sixty-nine years after Hubble's discovery, another certainty was demolished by observing a particular type of star, as we will now see. A first question that can be asked is whether the rate of expansion of the universe has been and will always be the same or whether it changes over time. From an intuitive point of view, the answer that can be given is the following. Let's consider a certain region of the universe containing galaxies. This region will tend to become larger due to the expansion, but the force of gravity will tend to slow down the expansion. Obviously, the larger the galaxies in the region, the less it will tend to expand. The rate of expansion will therefore be linked to the "battle" between the force of gravity, slowing down the galaxies motion, and expansion. From the theoretical point of view, the answer to the question we posed is linked to the equations of general relativity which describe how a system evolves in space and time. These equations tell us that the geometry of space-time is determined by the mass and energy contained in it. The greater this content, the greater the tendency for expansion to slow down. As we saw in Chapter 1, we have three possibilities. In the first case, the mass-energy is enough, and gravity would bring the system to a gradual deceleration until it reaches a maximum expansion and then re-collapse. The galaxies would begin to approach each other until the system implodes, resulting in what is called the *Big Crunch*, which is basically the opposite of the Big Bang. This can be expressed in terms of the expansion parameter, by writing $\Omega > 1$; that is, the universe will re-collapse if the density is greater than about 5 atoms per cubic meter.

If the density were less than or equal to the critical one, 5 atoms per cubic meter ($\Omega \leq 1$), the matter-energy would not be enough to stop the expansion and have the system re-collapse. Expansion would decelerate, but in any case the universe would expand forever.

From an observational point of view, it is possible to determine the rate of expansion from the initial era to today. To this aim, we have to remind ourselves that when we observe a distant object, we observe it as it was at the moment of light emission, not as it is today.

By measuring the speed of the recession of galaxies placed at different distances, we can determine the rate of the expansion of the universe in different eras.

To do this, we must determine both the distance of the object and its speed of recession with independent methods. The proportionality constant between speed and distance is the Hubble constant, which is precisely the rate of expansion of the universe. The main problem is the determination of the distance of the object. How do we do that? We can use *standard candles*, objects of known brightness. We could think of using *Cepheids*, but they are not good for our goal because they are visible up to some tenths of megaparsecs. This distance is not enough to establish what kind of geometry our universe has.

Moreover, at distances greater than 10 megaparsecs, the expansion rate varies with time and the distance depends not only on the Hubble law but also on the density and pressure of matter.

To determine the geometry of the universe, we must go up to distances of 1000 megaparsecs or larger. At these distances, the wavelengths are shifted by about 30% from the cosmological redshift, z, and at that time the universe was about 10 billion years old.

A particularly important type of standard candles, which are adequate for us, are *type Ia supernovae*. They originated from the explosion of carbon-oxygen white dwarfs (see section 5.5). Low-rotation white dwarfs have a maximum limit for their mass called the *Chandrasekhar limit*, equivalent to 1.44 solar masses. If the white dwarf is in a binary system, its mass can increase when the companion's mass falls on it. When the mass of the white dwarf approaches the Chandrasekhar limit, the temperature in its core can reach the values necessary for the fusion of carbon. In a few seconds from the beginning of the fusion, a large part of the mass of the star is subject to a thermonuclear reaction that releases a huge amount of energy capable of producing a violent explosion. In the explosion of type Ia supernovae (SnIa), the brightness grows very quickly and reaches an intrinsic brightness equal to that of a few billion suns. Furthermore, the intrinsic brightness is very similar in all explosions of this type because of the uniformity of the masses of the exploding white dwarfs. A brightness decay follows, and after a few hundred days, most of the SnIa disappear in the glow of the host galaxy. So, the mass of the white dwarfs at the moment of explosion is the same in all those stars, and the physics of the explosion that gives rise to the supernova is similar in the different white dwarfs. Consequently, the energy released and the characteristics of the explosion and of the supernova phenomenon are very similar in different supernovae, which thus constitute *standard candles*. For the sake of precision, there are differences between the brightness peaks of SnIa, but it is possible to correct these discrepancies using an empirical relationship between the peak of brightness and the rate of decline of brightness over time, found in 1993 by Mark Phillips. The faster the brightness decreases, the weaker is the supernova. In this way, SnIa behave as perfect standard candles.

They have the advantage over Cepheids to be much brighter and can therefore be observed from larger distances. The second quantity we need is their speed of recession, which can be determined by studying the redshift of the light they emit. Given the distances and velocities of the supernovae, we can evaluate the rate of expansion of the universe at different eras. However, there is a problem. SnIa are not frequent. A galaxy like ours produces one SnIa per century, and after the explosion their brightness is visible only for a few weeks. This means that it makes no sense to observe a distant galaxy and wait for an SnIa to explode, but many galaxies must be observed simultaneously in order to increase the probability of observing one or more supernovae. Thousands or millions of galaxies must be observed simultaneously.

In the 1990s, two projects were created with that purpose: the High-z Supernova Search Team led by Brian P. Schmidt and the Supernova Cosmology Project led by Saul Perlmutter. Perlmutter's project was born out of a "redirection" of a project on mass extinctions produced by astronomical causes. In 1984, the paleontologists David Raup and Jack Sepkoski claimed in an article the existence of a periodicity of 26 million years in mass extinctions. Two groups of astronomers, Whitmire and Jackson, and Davis, Hut, and Muller published hypotheses similar to those of the aforementioned paleontologists in the same year. According to them, mass extinctions originated from a dim star called Nemesis, from the name of a goddess of Greek mythology. This star would be a red or brown dwarf, which would disrupt the objects found in an immense sphere around the solar system, called the *Oort cloud*, made up of a huge number of comets. As a consequence, it would increase the number of comets moving towards the solar system with an increase in the probability of impact with

Earth. Perlmutter's supervisor was one of the astronomers mentioned, Muller. Perlmutter created an automatic search for weak stars. The project led to nothing, and in order not to throw away the work, it was converted into an automatic search for supernovae. The history of the other team is less peculiar. Observing about 1 million galaxies per night, the two teams discovered several SnIa located at different distances. By comparing these distances with the relative recession speeds, the two projects were able to calculate the change in the rate of the expansion of the universe from remote eras to today. An interesting thing is that the goal of the two teams was to measure the slowdown in expansion, according to the ideas prevailing at that time.

In this regard, there is a curious anecdote told by Lawrence Krauss in his book *The Universe from Nothing*. When he was at the Lawrence Berkley Observatory discussing, among other things, the hypothesis that the universe is flat, with 70% of matter in the form of energy from the vacuum, Perlmutter said to him, "We will prove that you are wrong." What they found was exactly the opposite: the universe had been expanding at an accelerated rate for 4–5 billion years, a result that earned Perlmutter, Schmidt, and Riess the Nobel Prize in 2011.

The first problem to be faced while determining the rate of expansion of the universe was that of having a catalog of supernovae at variable distances up to a few thousand megaparsecs. Phillips's supernovae were too close, so the two groups had to build the necessary catalog with painstaking work. To do this, they took five years, reaching supernova ends at 4000 megaparsecs.

Now that they had the required supernovae, they could know how much the universe had expanded from a given moment in the past, considering their distance and the redshift of light from distant galaxies that provides, among other things, their speed.

To establish the rate at which the expansion occurred, it is necessary to know the distance of the galaxies together with their redshift. If the rate at which the universe expands decreases over time, this means that expansion in the past was faster and therefore the universe took less time to reach its current state. As a result, the universe is younger than in an accelerated expansion universe. The light must therefore travel a shorter distance to reach us. If the expansion rate decreases, since a standard candle is brighter when it is closest to us, we will have supernovae brighter than those in a universe where expansion occurs at a constant rate. Now, what *astronomers observed was exactly the opposite: supernovae were less bright*. In other words, for a given redshift z, the supernovae were more distant than they would have been if the universe had expanded at a constant rate. So, the universe *was* accelerating.

Another fundamental point is that a universe made only of matter cannot produce accelerated expansion. In fact, gravity tends to slow down the expansion of the universe, and therefore there must be a new component that gives rise to a gravitational repulsion. To do this, a fluid with *negative pressure* is needed. The two groups reached the conclusion that a new component of the universe was needed to explain the data they had collected. In the analysis of the two groups, three different models were considered: two models made up only of matter, baryon, and dark matter — one with hyperbolic geometry $\Omega_m = 0.3$ and one with flat geometry $\Omega_m = 1$. Astronomers saw that supernovae were less bright than what was indicated by models in which the expansion of the universe slows down. To explain the data, they introduced a third model that, besides the material component, took into account the presence of a cosmological constant, interpretable as a negative pressure fluid in such a way that $\Omega_m = 0.3$ and $\Omega_\Lambda = 0.7$. In a cosmology of this type, the expansion is accelerated nowadays, while in the past, expansion was slower. Consequently, the universe took longer to reach its present state and it is older than in other cosmologies.

A second important point is that light must travel a greater distance to reach us, and consequently supernovae, like other sources, are less bright than in a universe with slowed-down expansion. The data of the work of the two groups is represented in Fig. 9.1 (top panel), which represents the distance (related to brightness) as a function of the redshift. At first sight, there are no great differences between cosmologies for a universe that slows down (the curve with the dashed line and the one with dots), namely the first two lines from below, and a universe that accelerates, namely the top solid curve. Looking carefully, one can see that more SnIa are distributed on the top line, corresponding to a universe that accelerates. In Fig. 9.1 (bottom panel) the differences between the models are clearly visible. The top line describes

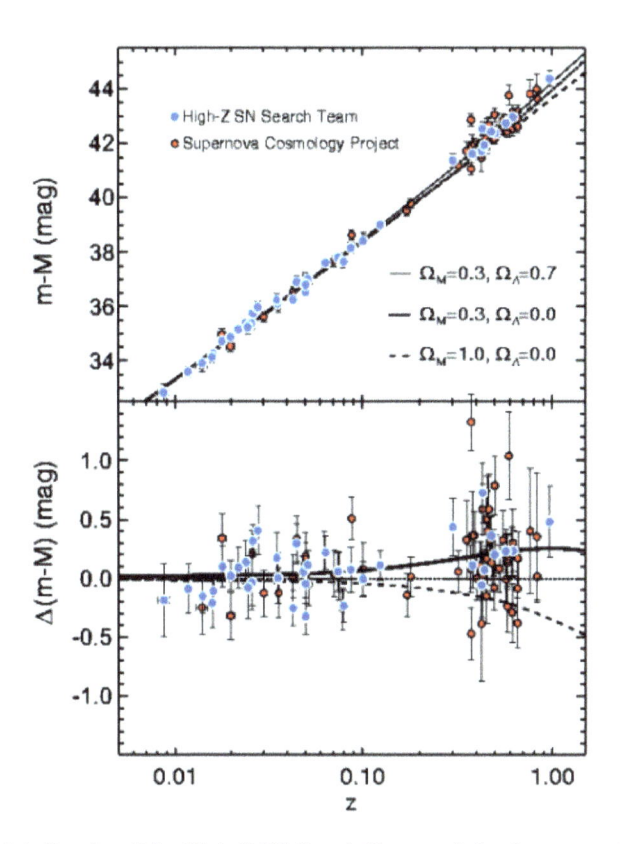

Fig. 9.1. Results of the High-Z SN Search Team and the Supernova Cosmology Project. The top plot is the redshift–magnitude relation (Hubble diagram) for high-redshift SNIa of the projects HZSNS and SCP. The bottom plot represents the top plot after subtracting the curve for $\Omega_{m,0} = 0.3$, and $\Omega_{\Lambda,0} = 0$. Credit: Perlmutter and Schmidt, *Supernovae and Gamma-Ray Bursters*, edited by K. Weiler, Lecture Notes in Physics, vol. 598, pp. 195–217.

the data better and corresponds to an accelerated, expanding universe.

The simplest form of this entity that accelerates the universe is the cosmological constant Λ.

Other confirmations on the accelerated expansion once again come from the CMB. The PLANCK satellite also measured the lens effect on CMB photons. At small angular scales, the image of the CMB is slightly distorted because of the deflection of light by the matter between us and the surface of the last scattering. The signal is dominated by the structures formed over the past 3 billion years. By combining these data with the CMB anisotropic spectra, information on the accelerated expansion of the universe is obtained, providing complementary evidence to that of supernovae.

9.2 Cosmological Constant and Dark Energy

So we need a component different from ordinary matter, which is called *dark energy*, and we know that $\Omega_\Lambda = 0.69$. The term is used for historical reasons. As we have already said, Einstein in 1917, in an article sanctioning the birth of modern cosmology, "Cosmological considerations on the General Theory of Relativity,"[1] introduced this constant to make the universe static. As shown by de Sitter, the cosmological constant does not compel the universe to be static. In fact, an empty universe with the cosmological constant expands exponentially. The cosmological constant in de Sitter's universe behaved as the source of a repulsive gravity, independent of the mass that was not present in his study. This gravity was uniform over time and constant in space, and as we will now see, another oddity of the cosmological constant gives rise to negative pressure.

The interesting thing is that the deprecated cosmological constant is able to explain the accelerated expansion of the universe.

How does it happen?

---

[1] Original title: "Kosmologische Betrachtungen zur allgemeinen Relativitätstheorie."

The equations of general relativity[2] consist in a left part that describes the geometry of space-time and a right part linked to the mass and energy contained in it. To explain the accelerated expansion, Λ is added to the right side of the equation and can be interpreted as a constant contribution to the energy density of the universe. It is an energy that fills the whole universe evenly. We do not know whether it really is a constant, nor its origin, so it is generally indicated by the term dark energy. The effect of this energy is that it produces an accelerated expansion of the universe. Λ produces a gravitational repulsion, an idea contrary to intuition because we know that gravity is always attractive. To understand why this happens, it is necessary to remember that, in general relativity, the gravitational forces are not only produced by matter and energy, but also by the pressure they exert. If the pressure is negative, a repulsion is produced.

Suppose we have a container with an ideal gas, having positive pressure, and compress it. To decrease the volume and compress it, we have to make an effort, and the gas accumulates energy. Suppose we have the same container full of dark energy. We should stress that this situation is not real, because dark energy extends everywhere and it cannot be enclosed in a container. Since the density of dark energy (made of the cosmological constant) is by definition constant, the energy in the container is proportional to the volume it occupies. By lowering the piston, as the volume decreases, the energy also decreases. Dark energy has the typical behavior of a system with negative pressure, or negative density. In a system with negative pressure, gravity behaves in the opposite way as usual. It acts like antigravity and produces a gravitational repulsion. Consequently, dark energy can explain accelerated expansion. If dark energy is the cosmological constant to explain the expansion in the terms observed in our flat universe, it is necessary to have $\Omega_\Lambda = 0.69$.

This changes the evolution of the universe. The appearance of the solutions of the equations of general relativity, as shown

---

[2] The general relativity equation with Λ is $R_{\mu\nu} - \frac{1}{2} R g_{\mu\nu} + \Lambda g_{\mu\nu} = \frac{8\pi G}{c^4} T_{\mu\nu}$.

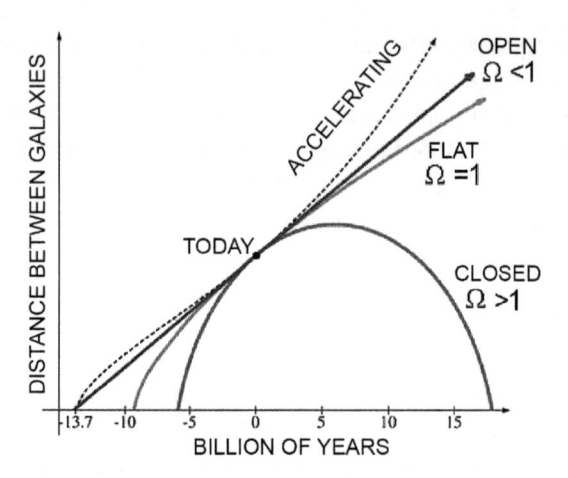

Fig. 9.2. Universe models. The dotted curve indicates an accelerated expanding model describing our universe.

in Fig. 9.2, change. From the Big Bang, the universe was subject to a decelerated expansion for about 5 billion years ago, after which the expansion accelerated (dashed curve).

In general relativity the cosmological constant is a simple term added to the field equations. One may wonder what the physical sense of this quantity is. One answer comes from quantum mechanics and *quantum field physics*.

The cosmological constant Λ is now interpreted as the energy of the quantum vacuum. But in this interpretation, there is a problem, the biggest in theoretical physics. In fact, vacuum is filled with all possible fields: the electroweak, QCD, Higgs, and in general bosonic and fermionic fields. While doing calculations with the theory of quantum fields, the same problem arises that we discussed — the hierarchy problem, in which virtual particles made large contributions to the mass of the Higgs boson. Again, each field makes very large (positive and negative) contributions, and certainly greater than the value of the dark energy coming from the observations. This problem was already known before 1998, and dark energy was assumed to be zero, implying a sort of magical cancellation among all the gigantic contributions that formed it, similar to what happens for the Higgs field. After the determination of the value of the cosmological constant, in 1998, and the discovery that it was not zero, it was found that the discrepancy between the observed and calculated values of the vacuum energy amounted to a factor 10^{120}. This huge discrepancy constitutes the *cosmological constant problem*.

This discrepancy is reduced in string theory to a factor $\sim 10^{50}$, still too large. Physicists hope that a future theory of quantum gravity, which would combine gravity with quantum mechanics, can solve the problem. Notice that the problem of the cosmological constant exists if Λ can be identified with the energy of the vacuum, but this is not certain.

It could (even if this is unlikely) be related to the Higgs field, or to other fields. Although Einstein, in 1931, eliminated the cosmological constant from his equations, it did not disappear from theoretical physics. The history of Λ is long and complex, as described in a 2017 article by O'Raifeartaigh.[3] It was not exhumed in 1998, because, after it was pulled out of the hat by the magician Einstein, it never wanted to go back.

9.3 The Geometry of Space

Apart from supernovae, there are other ways to show that our universe needs an additional component to explain observations. The geometry of space-time is determined by the amount of matter and energy it contains. In our discussion of CMB (Chapter 4) we have shown how its typical inhomogeneities have angular dimensions of about 1 degree, which implies that we are in a flat universe. The same result is obtained by comparing the images of the CMB map with simulations. Figure 9.3 compares the CMB map of the BOOMERANG experiment, already described in Chapter 2, with simulations, the three boxes under the BOOMERANG map. From the comparison it is clear that our universe has a flat geometry.

---

[3] https://arxiv.org/ftp/arxiv/papers/1711/1711.06890.pdf.

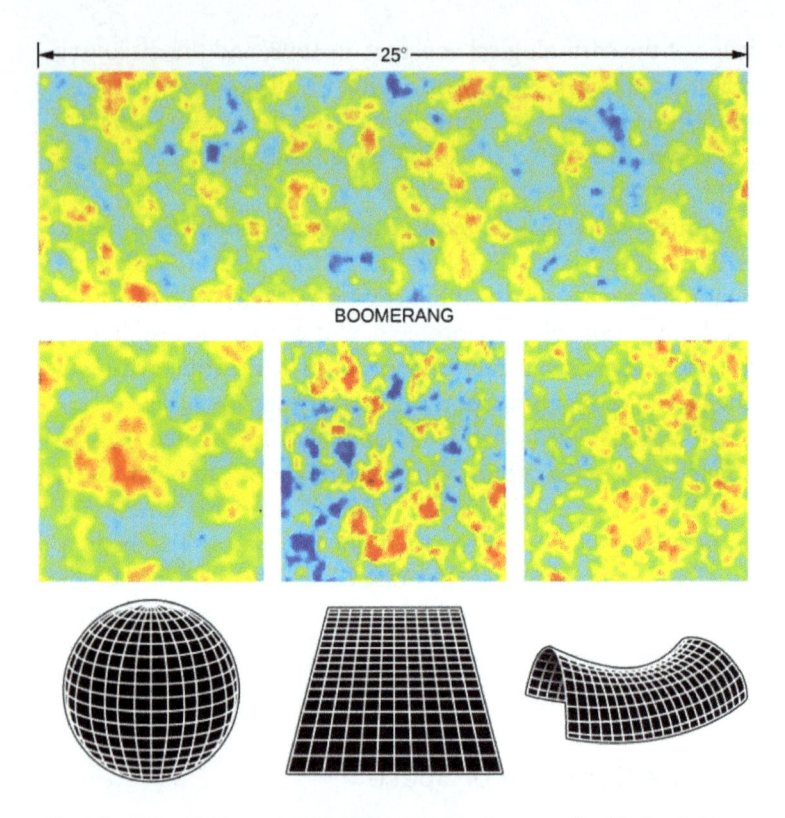

Fig. 9.3. Comparison of BOOMERANG data (top panel) with simulations (central panel) of a closed (left), flat (center), and open (right) universe. Credit: Modification of work by NASA.

In fact, the box that most resembles the BOOMERANG map is the central one, which corresponds precisely to a flat universe.

Referring to Friedman's solutions, shown in Fig. 9.2, our universe is represented by the black dotted curve.

As we already know, there is a parameter that gives indications on the material content of the universe in terms of critical density and therefore indicates the geometric structure and the future of the universe: the density parameter, $\Omega = \rho/\rho_c$. There is a value of the density parameter for each material component of the universe. For example, that relating to dark matter is $\Omega_{dm} = \rho_{dm}/\rho_c$. We already know that if $\Omega = 1$, the universe is flat. In order to get the value of Ω, the various forms of matter in the universe must be taken into account.

According to 2015 PLANCK satellite data, the contribution due to the baryonic matter Ω_{bar} is equal to 0.0486 (4.86%), that of dark matter, Ω_{dm}, is equal to 0.2589 (25.89%), and that (negligible) of radiation, Ω_{rad}, yields 0.00005 (0.005%), and Ω_v is 0.004 (0.4%) for neutrinos. Adding the indicated contributions we find that $\Omega = \rho/\rho_c = \Omega_{dm} + \Omega_{bar} + \Omega_{rad} + \Omega_v \approx 0.31$. As we have seen, however, our universe is flat, so $\Omega \approx 1$. This indicates that a matter-energy component of 0.69 (69%) is missing. As we know, that is dark energy. The sum of all contributions, including dark energy, provides $\Omega = \rho/\rho_c = \Omega_{dm} + \Omega_{bar} + \Omega_{rad} + \Omega_v + \Omega_\Lambda = 1.0023 \pm 0.0054$. This result tells us that

- the universe is flat and infinite in accordance with the inflation theory and that we only see a portion of it, the observable universe having a radius of approximately 46 billion light-years.
- the total energy of the universe is zero since the amount of positive energy in the form of matter (kinetic energy) is exactly canceled by the negative energy connected to gravity. This is one of the properties of inflation in which negative gravitational energy in the inflationary region balances itself with the inflaton field. In other words, inflation explains the cancellation of the energy of matter and gravitational energy on cosmological scales in accordance with the observations. The idea that the energy of vacuum is zero is, however, prior to that of inflation. Pascual Jordan first suggested the cancellation of the negative energy of the gravitational field and the positive energy of mass. As a consequence, in his opinion, a star could originate from quantum transitions without violating the conservation of energy. Gamow also discussed this idea with Einstein.
- the universe originated from a quantum fluctuation of the energy of the quantum vacuum. The idea was first published by Edward Tryon in 1973 in *Nature* and had many followers. So, despite the fact that many

philosophers have claimed that nothing is created out of nothing, according to inflation theory, the universe is "the ultimate free meal," in the words of Alan Guth. Everything can come from nowhere.

- the existence of a component of the universe with negative pressure has strong consequences on the evolution and future of the universe, which we will discuss below.

It is essential to be sure that dark energy exists and that the parameters which make up our universe are those obtained with the supernova method and with the CMB. Having a better understanding of dark energy has become one of the most important problems of today's cosmology. To understand the nature of dark energy it is necessary to use a variety of methods to measure acceleration.

A first method independent of supernovae is that of *baryon acoustic oscillations* (BAO). What is it about?

As we described in Chapter 2, the universe before recombination consisted of a plasma of baryonic matter and photons. The above-average density regions attracted matter and photons to produce pressure that opposed gravity.

This tug-of-war between gravity and pressure created oscillations, similar to sound waves. The regions with density larger than the average consisted of baryons, photons, and dark matter, and the pressure gave rise to spherical sound waves of photons and baryons moving outwards from the central region with higher density, consisting of dark matter, at a speed of 170,000 km/s. The acoustic waves traveled through the primordial plasma until it cooled and formed neutral atoms upon recombination. The photons were free to move, and they moved away at the speed of light while the baryons remained "frozen" on a spherical region. So, before vanishing, the spherical sound waves left an imprint of their existence on the matter of the universe (Fig. 9.4).

The baryons located in the spherical regions, and the dark matter in the center of the regions constituted regions of inhomogeneity that attracted matter, forming the galaxies. Sound waves traveled from the Big Bang to recombination

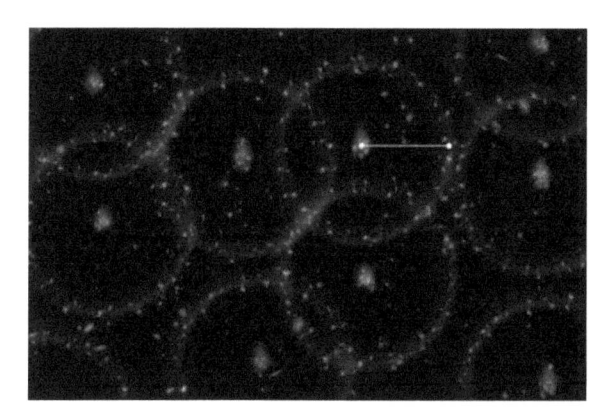

Fig. 9.4. Representation of the baryon spheres around the dark matter "lumps" (central points). Credit: BOSS collaboration.

for 380,000 years. The distance that the sound traveled in this period is indicated by the term *sound horizon*. Taking into account also the expansion of the universe, the spherical regions at recombination had a size of 450,000 light-years. The current size is 490 million light-years. As a consequence, we expect to see more pairs of galaxies separated by 490 million light-years, compared to a random distribution. This can be observed by studying the large-scale structures of the universe by means of surveys, such as the BOSS (Baryon Oscillation Spectroscopic Survey).

While supernovae SnIa supply standard candles, the BAOs supply a *standard ruler* of 490 million light-years for scale lengths in cosmology (Fig. 9.5), thanks to which it is possible to determine the distance.

BAOs can improve our knowledge of acceleration if the observations of the *sound horizon* today, using the way the galaxies are distributed, are compared to those of the sound horizon at the time of recombination, using the CMB. Therefore, BAOs provide a sort of known and standard-size ruler with which we can better understand the nature of acceleration, completely independent of supernovae (Fig. 9.5).

Another method for the study of acceleration and dark matter is related to the weak gravitational lensing of large-scale structures, also called *cosmic shear* (Fig. 9.6).

Fig. 9.5. Standard candles and standard rules.

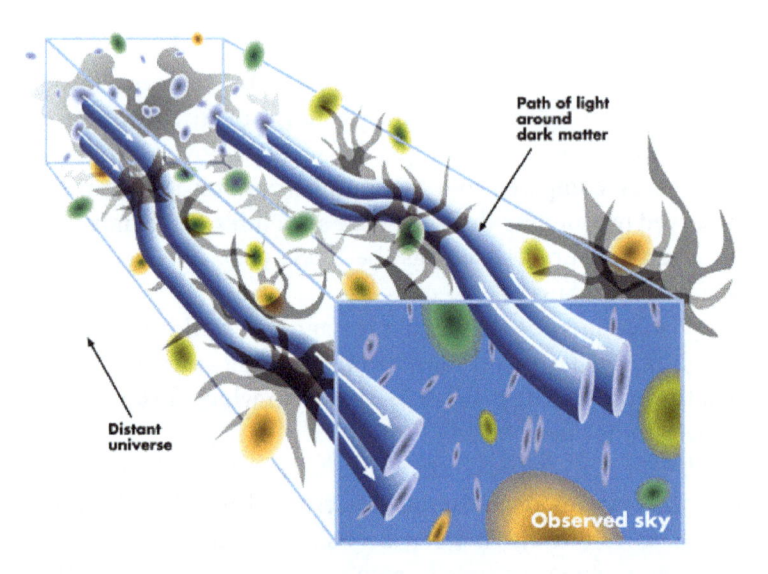

Fig. 9.6. Distortion of light from distant galaxies produced by dark matter between us and those galaxies. Credit: Bell Labs-Lucent Technologies.

Before reaching us, the light of faraway objects is deflected, and their images are amplified and deformed by the presence of the mass between these objects and us. The correlations between the deformations of the ellipticities of the galaxies can be used to reconstruct the mass projected along the line of sight.

This method complements that of the study of CMB anisotropies. In fact, since dark energy produces changes in the observables in terms of redshift, the CMB alone is insensitive to the parameters of dark energy models. The cosmic shear measurement in terms of the redshift can be used to test the impact of dark energy on the growth of structures and test different explanations and (gravitational) models of cosmic acceleration. In other words, the method is sensitive to the evolution of dark energy. Galactic sources at different distances suffer the lensing effect only from the masses in the foreground, allowing a *tomographic reconstruction.*

There are other methods used for the determination of cosmological parameters and the study of dark matter, but we will not discuss them. So far, every method converges with the results of the CMB, that is, the universe needs dark matter and energy to be explained.

The ΛCDM model or standard model of cosmology has been tested in thousands of articles, and apart from the problem of the cosmological constant, and discrepancies on small scales, it correctly predicts the observations. A doubt that can arise is that concordance, that is, internal consistency, may not be a sign of its exactness. For example, the Ptolemaic model with immense artifices approximately predicted the planetary orbits, even if it was not exact.

9.4 The Anthropic Principle, the Multiverse, and the Value of the Cosmological Constant

We discussed that Λ is fraught with a series of problems such as that of the enormous discrepancy between the values observed and those predicted by theory. Another is the *problem of coincidence*. Until about 5 billion years ago, the universe had not accelerated because dark energy was much less than it is today. Since then, it has increased so much that its value is similar to that of dark matter, and in the future, if dark energy is truly the cosmological constant, it will be larger than dark matter. The problem of coincidence can be stated as follows: why are the density of dark energy and that of dark matter comparable today? Either this could be a coincidence or there could be basic reasons for that. Several authors tried to predict the existence of dark energy and estimate the value of the cosmological constant. Few succeeded, apart from Steven Weinberg, famous for being, together with Abdus Salam and Sheldon Lee Glashow, the discoverer of the *electroweak theory* and for his contributions to the construction of the standard model. For these results he obtained the Nobel Prize in 1979. In a well-known article, he managed to establish the range of values in which the cosmological constant could be found. To do this he used an often criticized method, the *anthropic principle*. This term was coined by Brandon Carter in 1973. The idea of the principle is that the characteristics of the universe are tied to the existence of observers. Living beings can only live in particular physical environments. There are two forms of the principle, the *strong* and the *weak* one. This last, in Carter's words, is as follows:

> We must be prepared to take account of the fact that our location in the universe is *necessarily* privileged to the extent of being compatible with our existence as observers.

Whereas for the strong version, Carter says:

> The universe (and hence the fundamental parameters on which it depends) must be such as to admit the creation of observers within it at some stage. To paraphrase Descartes, *cogito ergo mundus talis est* .[4]

---

[4] I think, therefore the world is such [as it is].

This principle assumes a certain importance if it is applied to a multitude of physical systems with different characteristics. An example is that of *extrasolar planets*. If only our planetary system existed in the universe, it would be difficult to explain why it has all the characteristics to host life. This would lead us to think that "someone" has maneuvered the system so that it respects the characteristics necessary for birth, and evolution of life. After 1995, we have known that in our galaxy alone there are billions of planetary systems, each different from the other — some with stars bigger than the Sun and planets bigger than ours. There are myriads of combinations of parameters. Among the various statistical combinations, there are those that make life on Earth possible. There are billions of planets in which life does not exist and probably a handful in which it exists. Since we exist, we must find ourselves in one of these rare, or less rare, planets suitable for life. In order for life to exist in the universe, Λ must have particular values. For example, if the value of dark energy were too large, the accelerated expansion of the universe would have started too early or too intensely. This would not have allowed gravity to act, and structures, stars, planets, and therefore life formed. If dark energy had a negative value, the universe would have collapsed, and if the size of the energy had been very large, the universe would have collapsed before forming stars and planets and life. On the basis of these arguments, Weinberg established a range of acceptable values for the density of dark energy ρ_Λ: $-10\rho_m < \rho_\Lambda < 100\rho_m$, ρ_m being the density of ordinary and dark matter. Since dark and ordinary matter make up 31% of the total, we will have $\rho_\Lambda \approx 2.23\rho_m$, which is in the range predicted by Weinberg.

It must be remembered that to apply the concept of the cosmological principle, there must be many universes with different physical characteristics. As we said, there are huge contributions, both positive and negative to dark energy, and this usually leads to an excessively large value. So in most universes the magnitude of energy will be too large to allow the appearance of life. In a small number of them, these contributions will compensate and fall within the limit found by Weinberg and life will be possible. To make life on Earth possible, it is not enough that the cosmological constant has certain values, but the fundamental constants, the characteristics of the forces, must have particular values. Small variations of them would not produce the universe we experience. As Freeman Dyson said,

> The more I examine the universe and study the details of its architecture, the more evidence I find that the universe in some sense must have known we were coming.

If gravity were more intense, the universe would collapse immediately after the Big Bang. If it were weaker, it would expand rapidly. The result would be that neither stars nor planets nor anything else would form. Small variations in the constants of the strong or weak interaction would make the formation of chemical elements and therefore of atoms impossible.

So, in short, the existence of life leads us to think either that there was a great "architect" who thought of everything down to the smallest detail or that there is a huge number of universes. Some of these, for statistical reasons, will have the characteristics necessary to host life.

Ultimately, for the anthropic principle to work, there must be a large number of universes, that is, a *multiverse*, as this possibility is often indicated.

Generally, the word *universe* indicates everything that exists, so the term *multiverse* may seem like a contradiction. We could say that when we talk about a multiverse, we mean that the universe could have separate regions, not in causal contact. For those who live in a universe, it would be impossible to observe others, but possible to observe only their own. A simple way to get a multiverse is to consider a flat universe. It is infinite. However, because of the limited speed of light and expansion, we can only see a certain region, the observable universe. There are other possibilities for the existence of a multiverse.

For example, in the *many-worlds interpretation of quantum mechanics* proposed by Hugh Everett III in 1957, the universe divides into two different universes at each quantum measure. This implies the existence of a huge number of universes.

Another theory that leads to the multiverse is *eternal chaotic inflation*. Here we will mention this topic and discuss it in more detail in Appendix A.

In the primordial universe the energy of the vacuum was much greater than now and caused an exponential expansion, making its size grow enormously. The energy that originated this expansion was given by a field similar to the Higgs field, whose energy profile is similar to that of a Mexican hat (Fig. 9.7).

It is assumed that the field initially had an energy greater than the minimum. The field, called *inflaton*, was therefore not in a state of real vacuum but in a false vacuum. Figure 9.8 shows the potential of the field made by a flat area at the top and then a sort of valley that leads to a minimum. The potential is usually referred to as the *Mexican hat potential*, such as the one shown on the right of Fig. 9.7. The position of the balls show how the field evolves over time. The field, like all physical systems, tends to return to its natural state, to a minimum.

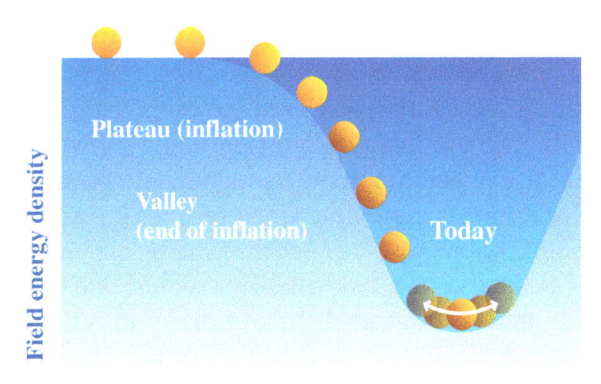

Fig. 9.7. Inflation. The balls represent the evolution of the inflationary field. Evolution of the inflationary field starting from a flat area (plateau), where inflation begins, the slow rolling along the valley, and the final oscillations of the field.

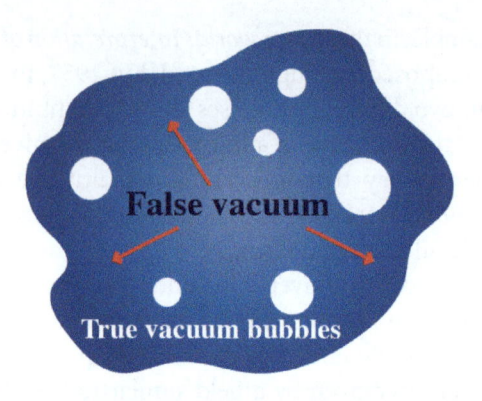

Fig. 9.8. Eternal inflation. The blue area is the expanding portion of the universe. The white areas represent the true vacuum bubbles and are universes not in contact and with different physical characteristics.

When the ball starts to move, the potential energy of the field turns into kinetic energy and the universe undergoes rapid expansion. Regions that were initially close, causally in contact, moved away very quickly, disconnecting from each other. When the field (the ball) descended to the base of the Mexican hat, the energy that the field had on top was released and particles, matter, and radiation were generated. The final density of matter that produces inflation is exactly equal to ρc, that is, $\Omega \approx 1$, equal to that observed. During the descent of the field, in the transition between the false vacuum to the true vacuum, a "bubble" of the true vacuum (white zone) is formed in the false vacuum (blue region) (Fig. 9.8). However, during the descent, the ball can arrive randomly in a generic point of the Mexican hat minimum. Consequently, in each region, the value of the energy of the vacuum will be different. Many bubbles of the true vacuum will form, not just one (Fig. 9.8), and each bubble corresponds to a universe. In other words, inflation ends at different moments at different points, and it is possible that in some regions it may repeat itself, generating an infinity of universes, a multiverse. This type of inflation is called *eternal inflation*.

As already stated, each bubble is a conventional universe and the set of bubbles is the multiverse. The bubbles that have formed expand wildly, at the speed of light. Despite this, they never manage to fill the space because it expands at a higher speed. In the version created by Linde, chaotic inflation creates an infinity of universes with different physical laws. This type of multiverse together with the anthropic principle can be used to solve the problem of the cosmological constant and the existence of life. In fact, if there are infinite universes with different laws and constants, one of the infinite universes will have the same characteristics as our.

The anthropic principle, although interesting and useful, is not held in good consideration by many scientists. Even Steven Weinberg, who, as we said, using it, managed to make quantitative predictions for the cosmological constant, criticizes it. The anthropic principle does not explain why the universe is as it is, why we exist, but on the contrary, it states that for the fact that we exist, the universe must be as it is. An unscientific perspective. Accepting it means renouncing to deepen the research and seeking — if the possibility exists — to understand nature.

There are other forms of a multiverse. In his book *The Hidden Reality: Parallel Universes and the Deep Laws of the Cosmos*, Brian Green mentions nine different types of multiverse. There would be the *landscape multiverse*, based on Calabi–Yau spaces, about which we talked in Chapter 7; the *quantum multiverse*, linked to the interpretation of Hugh Everett III quantum mechanics (mentioned a little earlier); the *brane multiverse*, based on M-theory, as already discussed, and each brane constituting a universe; the *simulated multiverse*, existing in very powerful computers, which would simulate entire universes — an idea similar to that of the movie *The Matrix*; the *holographic multiverse*, based on the *holographic principle*, which asserts that our three-dimensional universe is a kind of projection of a two-dimensional reality, like a hologram. Extending the idea to the multiverse, one gets the *holographic multiverse*. The dynamics of the inflationary multiverse would be contained in its surface. The list is not complete.

Max Tegmark has carried out a sort of classification of multiverses. There would be four different types of multiverse related to eternal inflation:

- Type I: universes similar to each other, but undetectable
- Type II: similar to case I but with different constants of physical laws and spatial dimensions.
- Type III: with the same characteristics as II, and linked to the many-worlds interpretation of quantum mechanics
- Type IV: multiverses with different forms of the laws of physics. Every mathematical structure has a correspondence in the physical world.

Although a myriad of articles have been written about the multiverse or *parallel universes*, as they are often called, proving their existence is at least very unlikely.

9.5 Quintessence, the New Ether

A non-zero cosmological constant justifies the accelerated expansion of the universe, but suffers from several problems. If it is linked to the energy of the vacuum, we know that there is a huge discrepancy between observed and calculated values. We do not know whether the fact that the density of dark energy and that of dark matter are similar today is a coincidence or indicates something fundamental. Given these and other problems, alternatives to cosmological constant have been proposed. One that enjoys particular interest is the *quintessence*, or *fifth element*, which in the Aristotelian worldview constituted the spheres and celestial bodies, from the lunar sky to that of the fixed stars. The history of ether began with the ancient Greeks; Plato, Aristotle, who gave it the name *ether*, and a systematic treatment. Unlike the four elements of the sublunar world, earth, water, air, fire, the ether was the essence of the celestial world, and precisely the fifth element. In the Middle Ages the term *ether* was replaced with *quintessence*, which was the constituent of the *philosopher's stone*, which in medieval beliefs was a substance with phenomenal properties, among which we remember the ability to confer immortality. Ether also had luck in the 19th century when it was postulated by supporters of the wave theory of light to explain its propagation in empty space. Its nonexistence was highlighted by Michelson and Morley's experiment in 1887 and opened the way for Einstein's theory of special relativity. Today, ether is a *scalar-type field*, introduced in 1988 by Ratra and Peebles, and extended by Caldwell and collaborators. The current name comes from the fact that the universe is made up of dark matter, baryons, photons, and neutrinos, and therefore dark energy constitutes the fifth element. Ether has a potential energy profile similar to that of Higgs and inflaton fields. Unlike the cosmological constant, which does not change with time, quintessence changes in time and perhaps also in space. Quintessence is usually represented by a perfect fluid with negative pressure, P, and density, ρ, which follows the law $P = w\rho$, called the *equation of state*, and w is called the *equation of state parameter*. It is generally negative, so as to have a negative pressure. When $w = -1$, we are dealing with the cosmological constant. There are models with $w < -1$, the *kinetic quintessence* (*k-essence*), with an untypical form of kinetic energy and the *phantom energy model* in which the kinetic energy is negative and the expansion is faster than that related to the cosmological constant. A universe dominated by phantom energy would lead to the *Big Rip* (of which we will talk later), an end of the universe similar to the one that Jack the Ripper reserved for his victims. As mentioned, there are similarities between the mechanism that produces expansion in the case of dark energy and that of inflation. It has also been tried, with not much success, to link quintessence and inflaton. However, in both cases, the evolution of the field is the same: it starts from a maximum of potential, the false vacuum, followed by a rolling phase towards the minimum of potential energy, the true vacuum, with the consequent acceleration and production of radiation and matter. The inflationary field and quintessence have other points in common. For example, they should not couple with ordinary matter particles, as this has never been observed. The shape of the potential is fundamental because it must reproduce cosmological observations. Quintessence has been proposed by several physicists as the *fifth force*. This force would modify the evolution of the cosmos. To date, we have only had knowledge of four forces, and to explain what is happening around us or in the solar system, we do not need a fifth force. Consequently, there must be a "chameleon-like" mechanism that shows the effects of this force only in certain conditions and hides them in others. For example, it must hide them in our laboratories and in the solar system, where observations are well described by Newton's equations. This hypothetical phenomenon is called *screening*, and there are several types of it (*Chameleon screening*, *Vainshtein screening*, and so on). However, all these theories have no experimental evidence, and the old and "battered" cosmological constant continues to be the most probable reason why the universe expands in an accelerated manner.

9.6 Do Dark Matter and Dark Energy Really Exist?

In a Russian fairy tale, there is talk of a beautiful princess, who on a day of terrible weather meets an ogre. At the end of the story, the author wonders: "Maybe it was all the other way round." Maybe the princess was terrible and the weather was beautiful, and so on. In other words, things can be different from how we imagine them. This could also be true of dark matter and dark energy. That is, given that decades of research on dark matter has given negative results, a spontaneous question is, are there other possibilities besides dark matter? In the same way, one may wonder if there are other possibilities to produce accelerated expansion beyond dark energy. Many scientists have dedicated themselves to giving a solution to this dilemma. We have seen that in the case of dark matter, the classic proofs of its existence, such as the rotation curves of galaxies, the mass discrepancy in the clusters, the lensing effect, are all based on the idea that known physics, and in particular the Newtonian mechanics and Einstein's theory of relativity, are correct.

In reality, the latter has been verified on certain scales, for example in the solar system, but not on all scales. In recent decades, various theories of *modified gravity* have been proposed with the aim of explaining the universe without the presence of dark matter and energy. Although the modified theories of gravity are not new in physics, the discovery of the acceleration of the universe has sparked interest in them. The basic idea is that although experiments are in agreement locally with general relativity, both in time and space, gravity may be different in the primordial universe or on large scales. One of the first to propose the idea that one could get rid of dark matter by changing gravity was Arrigo Finzi in 1963. His work was forgotten, and two decades later a similar idea was published by Mordehai Milgrom, the well-known MOND theory (Modified Newtonian dynamics), which changes Newton's laws at small accelerations, in the order of 10^{-8} cm/s^2. In this way it is possible to explain the rotation curves of galaxies without dark matter. MOND also explains several other things, but when you get to scales equal or larger than galaxy clusters, the problems start. In order for MOND to continue to work, it is necessary to reintroduce some dark matter in the form of sterile neutrinos. The situation is similar to that of the dog biting its tail. After the discovery of the accelerated expansion of the universe, in addition to the cosmological constant and the quintessence, a large number of theories of modified gravity have been proposed. This indicates that, especially in the case of dark energy, we have unclear ideas. In addition to the theories of modified gravity, another possibility is that on large scales, the universe is neither homogeneous nor isotropic, and that the Earth is in a huge empty region, and the rate of expansion would vary according to location, which could be interpreted as if it varied over time. Continuing to use the analogy of the rubber balloon with the cosmic expansion, if the rubber that constitutes it were different at different points, so as to vary the elasticity of the balloon, this would not inflate uniformly. The result would be that, depending on where we are on the balloon, we would see a variable rate of expansion.

In more technical terms, the universe is usually assumed to be homogeneous and isotropic, but in reality, on scales less than 100 Mpc, this is not true. Consequently, the inhomogeneity could influence the way the light propagates. Studies on a particular model of the universe (the Lemaître–Tolman–Bondi model) have shown that inhomogeneity mimics acceleration, that is, inhomogeneity is interpreted by us as accelerated expansion. This action of small-scale structures on the behavior of the universe on a large scale is called *backreaction*. Other doubts that the universe expands in an accelerated manner have recently been put forward by Jacques Colin and collaborators, who have shown that acceleration is not the same in all directions and that the acceleration deduced by supernovae would not be real but due to the fact that we are non-Copernican observers. Several scientists do not agree with this point.

Most scientists agree with the idea that we need dark matter and dark energy to explain our universe. Until direct evidence of the existence of dark matter and a greater understanding of dark energy are found, we must remain open to all possibilities. One hope for the near future is the *Euclid mission*, which could clarify these issues.

Chapter 10

END OF THE UNIVERSE

This is the way the world ends
Not with a bang but a whimper.

— Thomas Stearns Eliot

10.1 An Ancient Curiosity

The questions of how the universe originated and how it will end are probably as old as human civilization. Obviously, the answers to these questions has changed a lot with the evolution of populations. The questions about the origin and the end of the world are closely linked to that of the origin and the end of the individual — that is, life and death. The answers to these questions have been linked for thousands of years to mysticism and religions. With the evolution of societies, more philosophical and rational topics are sought. The eschatologies, the doctrines aimed at investigating the fate of the individual, of humankind and of the universe are myriad. Just to give some examples, the mythology of the Vikings and other Nordic peoples was completed with *Fimbulvetr*, a very cold winter of three years, followed by the *twilight of the gods*, the final struggle between the forces of order and those of chaos. This would lead to the end of the world, which would be reborn and repopulated. In Hindu cosmology, Brahma is the creator of the world and a day of Brahma, or Kalpa, lasting 4.32 billion years, is a measure of the cosmic cycles of evolution and involution of the universe. At the end of each day of Brahma, a night of Brahma occurs, during which the world is partially destroyed by fire, water, and wind. After 100 years of Brahma, Brahma dies and the universe is completely

Fig. 10.1. Death of the Sun and the Moon and fall of the stars. Credit: Cristoforo de Predis, 15th century.

destroyed and does not exist for the next 100 years of Brahma. So Brahma will be reborn, and with him, the world in a continuous cycle. This eschatology, as well as the Vikings', is similar to today's cyclical theories of the universe.

The end of time can also be the end of a civilization, as in the eschatology of the *Hopi Indians*, according to which the end of time coincided with the arrival of the Whites. During the end of time, the Earth will have been covered by an immense network (probably the telegraph), and iron snakes (railways and trains) and rivers of rock (highways) will have crossed the Earth. In their prophecies there was also talk of a "great place of sinking" and a great collapse, after which a blue star would appear and the Earth would become a desert of sand, rock, and icy water.

Unlike the vivid eschatologies of the peoples who lived on Earth, today's cosmology allows us to answer in a scientific, albeit not certain and unambiguous, way to the question relating to the end of the universe. In reality, there are several possibilities and these basically depend on the geometry of the universe and its future evolution, which depend on the relative proportions of dark matter and dark energy, the nature of which is not yet known.

Basically, the theories about the universe and its end are as follows:

- the universe is infinite in time, such as the steady-state theory, which has been disproved by various experimental data
- the universe has a beginning and an end
- the universe is cyclic: it takes birth, evolves, dies, and is reborn
- there is a multiverse: the universe is part of a space containing other universes

Let's take a look at these possibilities.

10.2 Thermal Death

The end of the universe depends on the geometry of space, and therefore on the content of dark matter and dark energy. Knowing the nature of dark energy, a certain answer could be given on the future of the universe. What we know today is that dark energy is accelerating the universe, and if things continue like this the future will face *thermal death*. However, if the behavior of dark energy changes, and for example becomes attractive, the future of the universe would be completely different (see section 10.4 on the Big Crunch).

In the case of thermal death, the universe would expand, the stars in the galaxies would go out, the black holes would evaporate, and the protons, if some theories are correct, would disintegrate. The universe would become an immense, almost empty, and cold place populated by photons, and with energy distributed evenly. This is the most accepted idea today about the end of the universe, even if there are conflicting points of view on this model, on the applicability of entropy and thermodynamics to the whole universe, and on making predictions based on the poor knowledge of entropy, gravitational fields, and the role of quantum effects.

The first ideas on the thermal death of the universe are linked to William Thomson, who proposed it in 1851. This end of the universe is linked to the second principle of thermodynamics because in an isolated system, entropy, namely disorder, increases irreversibly. If the universe lives long enough, it will reach a state with uniformly distributed energy, making the existence of processes connected to energy impossible, including life. A time scale of the end of the universe could be built. In about 3 billion years, the Great Magellanic Cloud, one of our small satellite galaxies, will collide with ours and a billion years later there will be a collision with Andromeda. Assuming that the solar system survives these events, in about 5 billion years the sun will run out of hydrogen and turn into a *red giant* expanding a hundred times. This will evaporate the oceans, melt the rocks, and so on. However, life, except the bacterial one, will have already disappeared. In about 600 million years, the level of carbon dioxide will drop below the level necessary for the production of photosynthesis. Almost all plants will die except those that use carbon-4 (C4) for photosynthesis. However, even these plants will not survive much, and with them, the animals. In about a billion years, the solar brightness will grow by about 10% and the Earth will be at the mercy of a *greenhouse effect* much greater than the current one. The oceans will slowly evaporate. Star formation will end in approximately 5×10^{10} years, and after 10^{12}–10^{14} years the stars will switch off. The cessation of star formation will initially produce only changes in the color of galaxies. After a few tens of millions of years from the formation of the last stars, those with masses greater than 8 solar masses will explode in the form of supernovae. The disappearance of these stars, which tend to be blue, will lead the galaxy to have a color tending to be yellow. No more heavy elements will form. Since the brightness of a star increases during its existence, for a few million years after stellar formation has ceased, the brightness will be kept constant. Then all the stars will form *fossils*, such as white dwarfs, neutron stars, and black holes, and the galaxy will be composed of these objects and brown dwarfs and planets. The temperatures of all objects will drop as they move around their orbits. The rotation around the orbit will not be eternal, since two objects that orbit one around the other emit *gravitational waves*. It is estimated that, because of gravitational radiation, most of the orbits of the stars in the galaxy will decay and they will be swallowed up by the central black hole. At that time there will only remain stellar fossils: white dwarfs, neutron stars, black holes. Over time, these objects will also be swallowed up by the central black hole, apart from some lucky objects that, through interactions with others, will acquire enough energy to escape from the central regions of the galaxy. The stars will separate from the galaxies in 10^{19} years and gravitational waves will cause the orbits to decay in 10^{20} years. In the case of the Earth–Sun system, this will happen in 10^{23} years. The galaxies will literally have evaporated. Even ordinary matter is not eternal. Leptons (e.g., electrons) are stable particles, neutrons decay in about 15 minutes, and whether protons decay is unknown. The future of the universe will depend on the stability of the protons. If these were not stable, the material of the universe would be transformed into iron and then into neutrons. Even protons joining electrons would form neutrons. In an unimaginably long time, all matter would be swallowed up by black holes, which will evaporate due to *Hawking radiation*. However, some *theories of Grand Unification* (GUT), such as the Georgi–Glashow model, foresee the decay of the proton. One of the possibilities of proton decay is the decay into a positron, a neutral pion, and 2 photons. The most recent experiment that studies this problem is Super-Kamiokande, containing 50,000 tons of ultrapure water made of 10^{34} protons and more than 11,000 photomultipliers, which reveal the blue light produced by the Cherenkov effect. According to the experiment, the *half-life*[1] of the proton must be greater than 10^{33} years. The results of Super-Kamiokande do not say that the proton does not decay but only gives a lower limit to its half-life. Maybe protons will decay in more than 10^{34} years, or maybe they

---

[1] The half-life is the time required for a given number of protons, or any other radioactive element, to reduce to one half.

won't decay. Decay products, such as positrons, will annihilate with electrons. Not even black holes will live forever but will slowly lose mass because of Hawking radiation and disappear in a sort of explosive photon emission.

A stellar black hole of 10 solar masses evaporates in 10^{70} years and the supermassive ones, at the centers of galaxies, with a mass of 1 million solar masses, do it in 10^{85} years, while one with 10^{11} solar masses, in 10^{100} years. So, all the energy and mass of the universe will be absorbed by black holes, which will eventually evaporate by emitting photons.

This last conclusion is valid if Hawking's radiation, which has never been observed, really exists. Like most physicists, we assume that it is a reality of our world. Black holes will absorb the mass and energy of the universe, and when they evaporate, only photons will remain, and perhaps electrons that have not disintegrated by meeting with positrons generated by proton decay.

In the classic Big Freeze scenario, the universe will be reduced to a cold and empty land in which photons, and perhaps electrons, roam.

Other scenarios have been added to this basic scheme that could occur after the Big Freeze, or before it occurs, by changing the final fate of the universe. One of the events related to the Big Freeze is the possibility of the *end of time*. To understand what this means, we need to have a more precise idea of what time is, and sincerely, we don't have it. The concept is extremely complex and for millennia it has been the subject of philosophical and scientific disquisitions. The statement of Augustine of Hippo is well known: "If no one asks me, I know; if I wish to explain it to one that asketh, I know not." Even modern science does not have univocal ideas about what time is. Ideas go from thinking it as an entity that pushes systems to evolve towards states of maximum entropy, that is, to the maximum disorder, to ideas such as those of Bryce DeWitt, Carlo Rovelli, and others, for whom time does not have to be part of a fundamental theory. This claim is based on the fact that time does not appear in the equations of the *quantum loop theory*, which attempts to unify general relativity and quantum field theory. On a more trivial level, if we take up Galileo's anecdote that measures the time of the pendulum's oscillations with the beat of one's wrist (Chapter 6), we realize that in reality we never directly measure time. We measure time between the oscillations by comparing it with the heartbeat, but to find out if the beat is truly regular, we compare it with a clock, another pendulum. The situation is similar to that of the dog biting its tail.

Trying to give an operational definition of time, one can consider it as a quantity that is measured with appropriate instruments, such as watches, intended not only as mechanical instruments but also as generic cyclic systems, such as the Earth–Sun system. In other words, time-measuring instruments, such as clocks, are based on the comparison between a movement in space, for example the Earth's rotation, and sample movements (mechanical, hydraulic, electronic), with sufficient precision and reproducibility.

Time could be identified with movement. This was also the point of view of Aristotle, who hypothesized that time was movement since it can be measured by the movement of the Sun in the sky or sand in an hourglass.

If the instruments of the universe suitable for measuring time disappear, time would itself disappear, together with the dimensions and distances, as Alexei Filippenko and the Cambridge cosmologist Gary Gibbons think. Ultimately, time had a origin, with the Big Bang, and could therefore end, leaving everything of which the universe will be made up in a frozen state, as in a still image.

The imagination of physicists is very prolific, and there are therefore other possibilities that avoid *thermal death* or ideas that allow the universe to go beyond thermal death. If the vacuum of which the universe is made were a false vacuum, after enormous times (10^{2500} years), the false vacuum could decay into a true vacuum and the universe could experience again a Big Bang. This scenario is the same as those of universes dominated by the false vacuum that can lead to a sudden disappearance of the universe in a *Big Slurp*, which we will see in the next section.

In summary, the classic thermal death of the universe requires that we arrive at a cold universe containing only photons, but some ideas would assume that a new universe, even cyclical, or multiverses can originate from it.

10.3 Big Slurp

We have seen that quantum vacuum is the minimum energy state of a physical system. However, a system can be trapped in a state with energy greater than the minimum, that is, it can be in a false vacuum. The system is therefore not stable. If the potential of the Higgs field is like that of Fig. 10.2 (left panel), the field will evolve to a minimum and stop. If the situation is like that of Fig. 10.2 (right panel), the yellow line shows that there is a second minimum with lower energy. So the field could evolve, in some way (e.g., through *quantum tunneling*[2]) and reach this minimum. The transition from a vacuum of higher energy to one with lower energy is called *vacuum decay*. If this happens at any point in the universe, the bubble of the new (true) vacuum would expand at the speed of light, changing the characteristics of our universe or

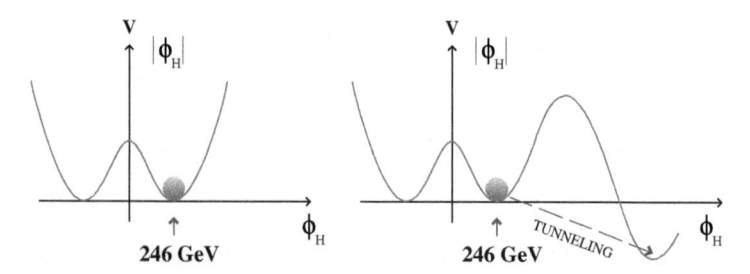

Fig. 10.2. Left: Higgs field in the minimum of the potential. Right: If there is another minimum in the potential of the Higgs field, the Higgs field can move to the lower-energy state.

---

[2] The tunnel effect is an effect of quantum mechanics that allows a microscopic object of the quantum world to be in a state that classical mechanics prevents. For example, in the classical world a bullet that does not have enough energy will not be able to cross a wall, while a particle in the quantum world can.

destroying everything. That this happens is unlikely because a huge amount of energy is needed. Furthermore, the time for this to happen is very long. According to a 2018 study by Andreassen and collaborators, the time needed would be greater than 10^{58} years.

This mode of ending the new universe is referred to as *big slurp*.

After determining the mass of the Higgs, it was clear that our universe is in a *metastable state* (Fig. 10.3), that is, halfway between stable and unstable.

This means that the Big Slurp scenario is one of the possible ends of our universe.

What will happen after the Big Slurp is completely speculative. One possibility, extending some of Caroll's 2004 results, is that, if the field that generated inflation still existed in the vacuum, it could reproduce itself.

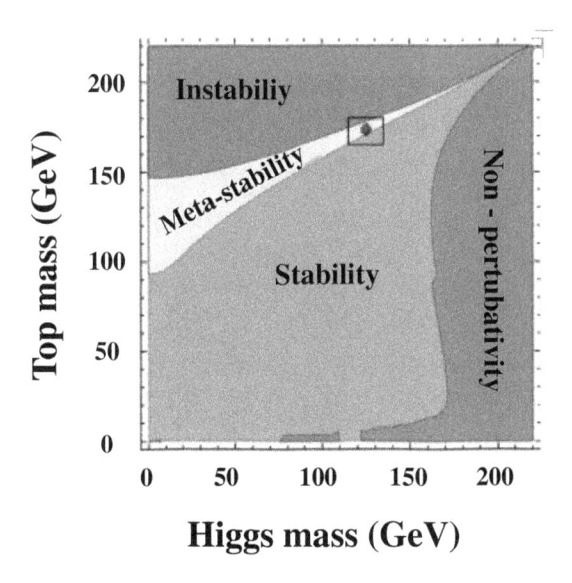

Fig. 10.3. Diagram of the stability of the universe based on the mass of the top quark (vertical axis) and that of the Higgs boson (horizontal axis). The values of the masses of the two particles tell us that the universe is located in a metastable area, that is, of precarious balance. Credit: Degrassi *et al.*, *J. High Energ. Phys.* 98 (2012). https://doi.org/10.1007/JHEP08(2012)098.

10.4 Big Rip or Big Crunch?

Another scenario that has a similar ending to the Big Freeze is the Big Rip. The end of the universe basically depends on the unknown nature of dark energy. If it is not a cosmological constant, but it has a dynamic nature, that is, it can change over time as in quintessential models, several endings are possible.

In 1993 Robert R. Caldwell, Marc Kamionkowski, and Nevin N. Weinberg proposed another idea for the end of a universe dominated by dark energy: the Big Rip. This would happen in the case of a particular form of dark energy called *phantom energy*, characterized by a *state parameter*, w, that is, the ratio between pressure and density, lower than -1 ($w = \frac{p}{\rho} < -1$). In this case, a much more accelerated expansion would occur than in the case of a cosmological constant, as shown in Fig. 10.4.

The expansion of space would be so rapid, owing to the repulsion produced by dark energy, that no force in nature, not

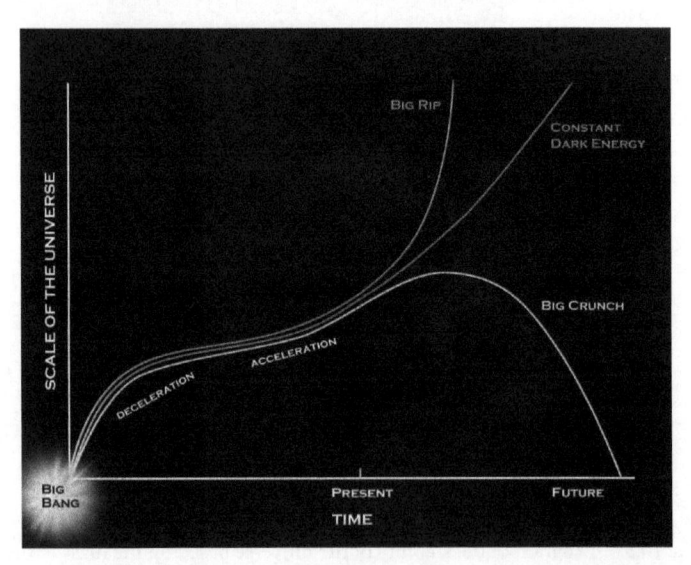

Fig. 10.4. Possible destinies of the universe. Accelerated expansion continues to be as it is today if dark energy is constant, or the expansion may be very accelerated because of the growth of dark energy, leading to the Big Rip. The last possibility is that of contraction and re-collapse (Big Crunch). Credit: NASA/CXC/M. Weiss.

even strong interaction, could counteract it. As a consequence, all the structures that make up the universe, starting from galaxies, stars, planets up to molecules, atoms, hadrons, and so on, would be broken down into elementary particles: photons, leptons, and protons (if the latter were not subject to decay). This event could happen, according to some theorists, in 20 billion years.

The final result of the Big Rip would be similar to that of the Big Freeze, but it would happen in much shorter times, that is, the structures would be destroyed in less time than what happens in the case of thermal death. The subsequent events would be similar to what was described in the case of the Big Freeze with the end of time and space. However, current studies on dark energy lead us to think that it probably does not have the characteristics of phantom energy. For these reasons, the universe is more likely to be dominated by a dark cosmological or quintessential constant energy. If so, the universe would end up in a Big Freeze. The end of the universe would be something similar to what T. S. Eliot writes in his poem "The Hollow Man": "This is the way the world ends / Not with a bang but a whimper."[3]

The case $w = -1$, corresponds to the case in which dark energy is the cosmological constant. Again, the expansion will be endless. The opposite possibility to that of the Big Rip is that the universe expands to a maximum and then re-collapses in the Big Crunch. This can happen, for example, if the spatial geometry is closed: $\Omega > 1$. Today we know that the universe does not satisfy this condition, so if it were made up only of matter, it would tend to expand forever. However, the existence of dark energy must be taken into account. In the event that the

---

[3] A model linked to the Big Rip is that of Lauris Baum and Paul H. Frampton. According to this model, an instant before the Big Rip (10^{-27} s), the space would be divided into a large number of subspaces, independent volumes, related to observable universes. Such mini-universes containing neither matter nor energy nor entropy would contract and each would give rise to a new Big Bang and a new universe. A multitude of universes, or a multiverse, would therefore be created, as in the scenario of chaotic inflation and string theory.

nature of dark energy totally changes and becomes attractive ($w > 0$), the universe would be subject to the Big Crunch, as shown in Fig. 10.4. One possibility is that the Big Crunch will be followed by a Big Bang generating a cyclic universe, which in standard cosmology, however, has problems with entropy, as we will see in the next section.

10.5 The Big Bounce

After solving the equations of general relativity, Friedman in 1922 found three solutions, one of which described the universe as cyclic, that is, consisting of a succession of Big Bangs and Big Crunches, similar to the situation of a ball moving away from the ground up to a maximum height, followed by a fall to the ground and a rebound (Big Bounce).

This was also one of the models designed by R. Tolman. Each bounce would have generated a singularity that cannot be described by general relativity. One of the problems of the model is to explain what happens to entropy, that is, the degree of disorder, in the various phases. The initial entropy of the universe, at the first Big Bang, was low and grew as its size

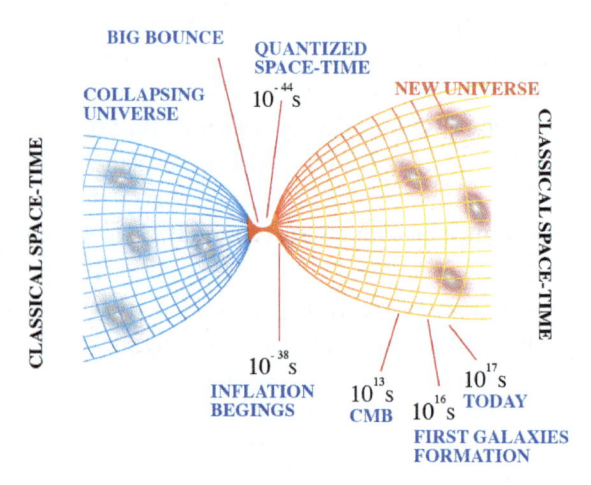

Fig. 10.5. Big Bounce. The shrinking universe on the left reaches a minimum size followed by a new expansion.

grew. In the re-collapse phase, entropy does not reverse its course, but as pointed out by Tolman in 1931, it will continue to grow. Subsequent to each cycle, there will be an increase in entropy with a consequent increase in the maximum size and time between successive cycles (Fig. 10.6).

With the discovery of the expansion of the universe, Tolman's universe was discarded. By adding a cosmological constant, which describes accelerated expansion according to the equations of general relativity, cyclicity is eliminated, as shown in Fig. 10.7.

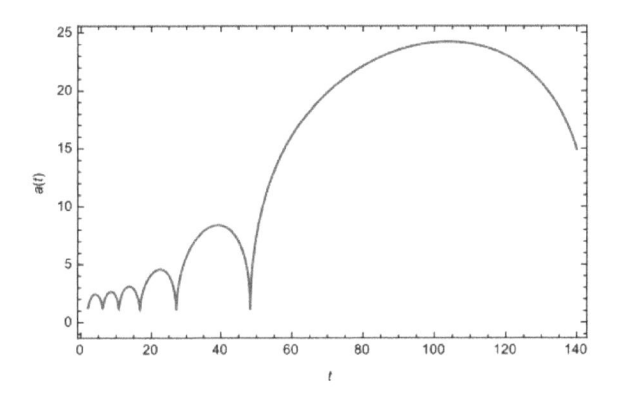

Fig. 10.6. Evolution of a closed universe due to the increase in entropy. Credit: Barrow and Ganguly, "The Shape of Bouncing Universes" (https://arxiv.org/pdf/1705.06647.pdf).

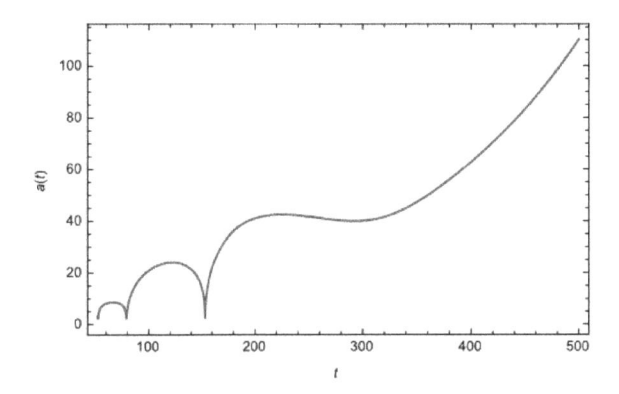

Fig. 10.7. Evolution of a universe with a cosmological constant because of the increase in entropy. Credit: Barrow and Ganguly, "The Shape of Bouncing Universes" (https://arxiv.org/pdf/1705.06647.pdf).

However, there remains conflicting theses. Frampton in 2015 showed that entropy can be eliminated from one cycle to another. This point of view is similar to other studies related to string theory. If we want a cyclic universe, it is necessary to find a way to decrease entropy. In 1999 Paul Steinhardt and Neil Turok believed they had found a possible solution inspired by string theory.[4] Steinhardt, who had been a major contributor to inflation theory, disagreed with the idea of the continuous birth of universes predicted by the theory. This is why he thought of finding an alternative to inflation. Steinhardt and Turok imagined two brane universes, each of which consisted of nine spatial dimensions. Only three of the nine dimensions were visible and the others were compactified. The universes were immersed in a ten-dimensional bulk. The two branes were linked together by a closed string, which carries gravity, and separated by an extra dimension that could contract periodically (see Fig. 7.6). Since the gravitational force, and the particle that mediates the interaction, the graviton, propagates between two branes, and given the contraction of the extra dimension, the two brane universes attract each other, giving rise to a collision called the Big Splat.

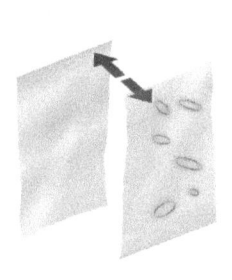

Big Bang from branes collision.

Branes move away because of freed energy.

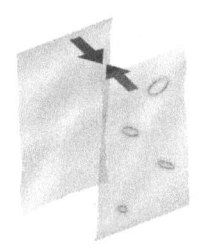

Branes approach because of extra-dimension contraction.

Fig. 10.8. Ekpyrotic universe.

---

[4] More precisely, the M-theory, which conveys the five string theories in a single theory.

This event precedes the Big Bang and is a fundamental phase in the formation of a cyclic universe. This universe was called the *ekpyrotic universe*.[5]

According to the two authors, the entropy problem was solved since only the extra dimension was contracted and not the branes, which continued to expand. The entropy created in the branes spread and was never concentrated. In this model, dark matter and dark energy were also explained.

Dark matter would be the manifestation of the gravity of another brane and the ordinary matter it contains. The collision between the branes would produce enormous energy and heat which, generating a gigantic explosion, would give rise to a new universe in accelerated expansion. The latter would have originated from the residual dark energy of the collision. After the Big Splat, the branes would move away and return to approach each other, because of the gravity and the cyclic contraction of the extra dimension, giving rise to a new cycle. Unlike the classic Big Bang and inflation model, the ekpyrotic model completely eliminates the singularity. Dark energy is a kind of extra gravity that, among other things, keeps branes aligned and stabilizes them. The branes are not rigid, but they are like sheets in the wind, so the collision takes place at different points and instants. At the points where it occurs first, the matter and energy produced by the collision will be significantly reduced because of the expansion. In the places where the collision occurs later, the inhomogeneities that will give rise to the galaxies originate.

So the initial inhomogeneity from which the galaxies originated does not need inflation to be explained; it is an integral part of the model.

Is reality better described by inflation or by the previously described model? To know this, it is necessary to make comparisons between the predictions of the two theories and the observations.

---

[5] This term can be translated as "transformation into fire" or "coming out of fire," a term that in Stoic philosophy indicated the moment in which the world was cyclically created and destroyed.

Inflation predicts that there are signs of accelerated expansion on the CMB: the so-called B modes mentioned in section 4.3, due to the generation of gravitational waves during inflation. The ekpyrotic model does not foresee them. In 2014, the BICEP-2 experiment seemed to have found this effect, confirming inflation. Unfortunately, this result was quickly disproved and therefore other data are requested in order to understand which one of the two models (ekpyrotic universe, or inflation) is correct. It must be said that Linde and collaborators have shown that the ekpyrotic universe has a series of inconsistencies with M-theory. For this reason, Steinhardt and Turok have added more precise details and have also changed the model name to *phoenix universe*. Studies on this model continued until two different possible models were proposed, one based on quantum mechanics, proposed by Turok and Gielen, and a classic model proposed by Steinhardt. In both, the singularity is avoided by obtaining the Big Bounce.

The Big Bounce has also been brought to the forefront by *loop quantum gravity theory*.[6] Taking into account quantum mechanics, a contracting universe cannot be crushed beyond a certain limit, as an electron can only approach the nucleus up to a certain distance. It is as if quantum mechanics introduces a repulsive force that produces a big bounce.

Finally, another type of cyclic universe was proposed by Roger Penrose, the so-called *conformal cyclic cosmology*. According to Roger Penrose, in a universe that has achieved thermal death, the microscopic level could have influences on the macroscopic one, causing the "dead" universe to give rise to a new Big Bang. The entropy problem is solved by assuming that information and therefore also entropy is lost in the final evaporation of the huge black holes that dominate the universe after 10^{100} years. In addition, the model predicts that the collision between black holes should produce gravitational waves visible as concentric rings on the CMB. According to Penrose and Vahe Gurzadyan, these rings are visible in the CMB map of the WMAP satellite.

---

[6] This theory aims to build a single theory starting from general relativity and quantum mechanics.

According to Steven Weinberg, "the oscillating universe ... pleasantly avoids the problem of Genesis" and would be attractive for this reason.

Unlike the Big Bang, which has been deeply tested, the previously described models have not received any confirmation from the observations, and as Richard Feynman said, "It doesn't matter how beautiful your theory is, it doesn't matter how smart you are. If it doesn't agree with experiment, it's wrong."

10.6 Multiverses

In section 9.4, we saw that Hugh Everett's interpretation of quantum mechanics, Andrej Linde's chaotic inflation or bubble theory, string theory, and so on, predict the existence of the multiverse. Even a flat universe is compatible with the idea of multiverse, given that from our universe we can only observe a part of the whole universe. The multiverse summarizes in itself all the possible endings that we have previously described. In other words, a multiverse is an infinite set of universes with (possibly) different physical laws that admit a multitude of different endings for their existence. Although there are many scientists who oppose the idea, the fact that both string theory and some inflationary models and flat universes converge on the multiverse idea could be an indication that the idea should be taken more seriously (see Appendix A). Among other things, putting together the idea of multiverse with the anthropic principle would eliminate the need for some "entity" to have set all the necessary parameters for the universe to form.

In any case, even if other universes exist, we are left with the problem of finding observational evidences of their existence, and this is highly improbable.

10.7 The End of Cosmology

We now ask ourselves how the beings who live in the universe will see it in a remote future. This is why we assume that dark energy continues to have the properties it has now. In this case from our galaxy we will see the galaxies continue to move away with increasing speed until it exceeds the speed of light. Strange as it may seem, this is not contrary to the predictions of relativity, which tells us that the maximum possible speed is that of light. In our case, the space is expanding, as space is created between the galaxies, at a speed higher than that of light, and this is not in contradiction with relativity. When the relative velocity of galaxies exceeds that of light, we can no longer see them. In fact, the light that starts from them will move at a speed lower than that of the galaxies moving away, which will therefore be invisible and leave our horizon. Obviously, they will not disappear suddenly from our horizon, but their light will be subject to the redshift. We will first see them become increasingly red, then their light will move towards infrared, microwave, radio wave, and so on. The galaxies of the local group, which would be our galactic neighborhood consisting of about 70 galaxies in a radius of about 1.5 Mpc, will continue to be visible because they are gravitationally linked to us. The time for this to happen is about ten times the present age of the universe. In about 100 billion years, our Sun will have already passed through the red giant phase, generated a white dwarf that will have cooled into a black dwarf. Our galaxy after colliding with Andromeda and other galaxies will be different. There will be stars and perhaps civilizations on some planet linked to them. These populations will not see the universe as we see it today. Aside from the galaxies in their surroundings, they will see nothing and will conclude like 19th-century astronomers that the universe is static and made up only of our own and neighboring galaxies. They won't even understand that the universe is expanding.

The three fundamental pillars of the Big Bang will no longer be available. The expansion of the universe observed by Hubble will no longer be visible as only the galaxies of the local group can be observed which, following various collisions, will form a single huge galaxy. We would no longer have evidence of the existence of dark energy, although by far it will dominate the density of the other components of the universe, because we will observe neither expansion nor acceleration. In a universe a few hundred times the current one, given that the temperature decreases with the size of the universe, the CMB temperature will be a factor 100 times less, and its intensity hundreds of millions of times lower. We would still have the third test: the abundance of light elements. At nucleosynthesis, the amount of hydrogen and helium was about 75% and 25%, not much different from today. The predictions of the primordial nucleosynthesis were verified by observations. Over time, the amount of helium and metals[7] will be much greater than today and there will be no need to assume a Big Bang to agree with the observations.

The latter description of the future of the universe, which is nothing but the scenario of thermal death, is based on the hypothesis that dark energy does not change its behavior. If not, we would have completely different scenarios, as described in the previous sections. If dark energy retains its nature, our possible descendants should invent a way to describe an apparently static universe, as Einstein already erroneously did. The future described in this chapter is the possible future, a future that could change, as a human being does according to his or her choices. As Mark Twain said, "It is difficult to make predictions, especially about the future."

---

[7] In astronomy, all the elements heavier than hydrogen and helium are referred to as metals.

APPENDIX A: INFLATION

Inflation is a prequel to the conventional Big Bang theory.

—Alan Guth

As we described in Chapter 1, just around 10^{-35} s after the origin of the universe, the universe underwent an exponential expansion, produced by a field called *inflaton*, leading it from microscopic dimensions to macroscopic dimensions, increasing its radius by a factor 10^{26}–10^{30}. The theory was initially proposed in 1979 by Alexei Starobinsky and a few years later by Alan Guth. It was introduced to solve some of the problems of cosmology, such as

- the problem of the horizon, namely the problem of explaining how regions of the universe that are not causally connected (that is, larger than the size that light could travel since the Big Bang), gave rise to a homogeneous and isotropic universe as seen in the CMB
- the problem of flatness, that is, the fact that the universe is described by a flat geometry
- the problem of topological defects, such as magnetic monopoles, which should, according standard cosmology, be present in the universe but are not.

There are several theories of inflation. The first one, Guth's inflation, assumes that the universe was initially in a state of false vacuum and that it transited to true vacuum by means of quantum tunneling (Fig. A1).

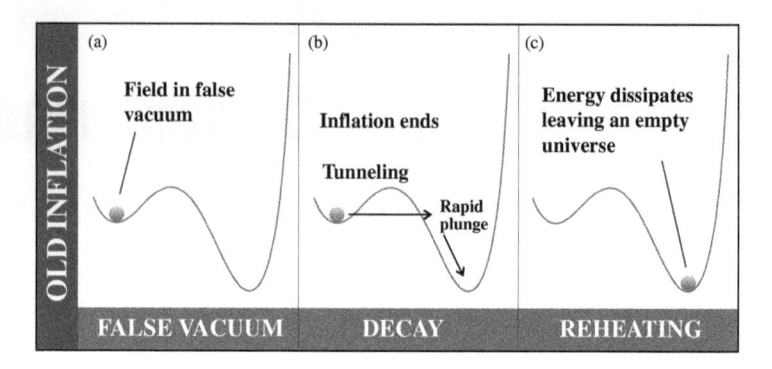

(a)	(b)	(c)
OLD INFLATION Field in false vacuum	Inflation ends / Tunneling / Rapid plunge	Energy dissipates leaving an empty universe
FALSE VACUUM	DECAY	REHEATING

Fig. A1. Guth's inflation.

This would have caused the appearance of "bubbles" of energy (bubbles of true vacuum in false vacuum), which would have had to collide and form a universe. Unfortunately, inflationary expansion was so fast that despite expanding at the speed of light, those bubbles could not coalescence. Then, the model had problems determining the end of inflation.

The model was replaced by another one (Linde and Steinhardt model) in which inflation was slower. The basic idea is that the universe was dominated by a *scalar field*, dubbed *inflaton*, similar to the Higgs field or to that of quintessence, and that the universe was initially in a state of the false vacuum of that field. That field was the only field then present — distributed throughout space (similarly to the Higgs field) and with an energy profile (potential energy) similar to that of the Higgs field. The starting part of the profile of the field, the false vacuum region, is flat as can be seen in Fig. A2.

From the state of false vacuum, that is, from the maximum of the curve, the field would have moved, slowly rolling (Fig. A2) towards the true vacuum, the minimum of the curve, that is, the region of minimum energy, characterized by repulsive gravity.

This would cause a rapid expansion of space. The field would then begin its rolling phase towards the true vacuum (minimum of the curve). The energy at the top, in the false vacuum, is greater than that in the zone of minimum, true

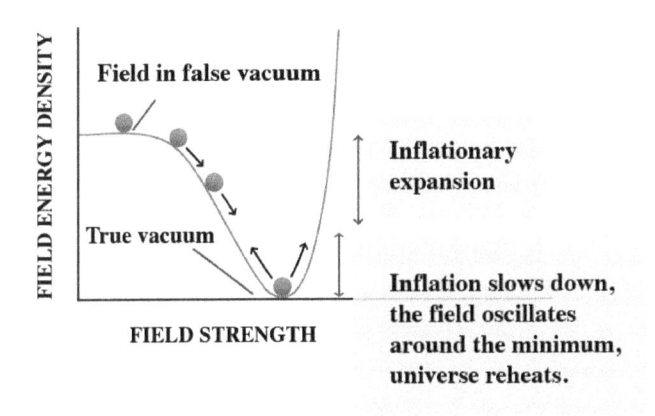

Fig. A2. Evolution of the inflationary field.

vacuum. When the field reaches the minimum of potential, in the state of true vacuum, it begins to swing (oscillate) around it, like a ball dropped along a slope with the shape of the potential profile shown. In this way, it would begin to dissipate its energy, located at the top of the curve. The release of energy, according to the rules of quantum mechanics, produces fields and particles in the so-called *re-heating phase*. Today's particles, therefore, originated at this stage of the evolution of inflation (Fig. A2). When the field reached the minimum of potential, that is, the real vacuum, the rate of expansion would slow down to today's rate.

In summary, the universe was full of the inflationary field that was in a state of false vacuum. Quantum fluctuations brought the universe out of the false vacuum state and pushed it towards the real vacuum state. This initially happened slowly, in the *slow rolling phase*, in which the universe enormously expanded from dimensions of the order of 10^{-28} m up to a few tens of centimeters over the time that the field took to reach the minimum of potential. This dimension is the observable universe. According to Guth, the size of the universe may be 10^{26} times larger than that of the observable universe, but there are much larger and smaller estimates. At this point, fields and particles were generated and the region in which inflation took place began to expand at today's rate.

Like all fields, the *inflaton* is subject to quantum fluctuations. Quantum fluctuations cause the potential of inflation to be subject to uncertainty, which means that inflation cannot end everywhere at the same time, because "fluctuations give different values to the field, even in two very close regions. So inflation would end at different times, and at different points. At points where quantum fluctuations exceed a certain threshold, inflation will occur, while for points where fluctuations are below the threshold, it would stop. A bubble of energy and matter would be created, which would give rise to a new universe. Once inflation begins, there will always be an exponentially expanding region that will give rise to another universe. This is one of the variants of inflation, the *eternal inflation*, in which expansion continues forever in different regions of the universe. Andrei Linde proposed a theory based on eternal inflation, called chaotic *inflation* or *bubble theory* (Fig. A3). In this model, once inflation begins, it will generate an infinity of universes, a multiverse, all with different physical laws.

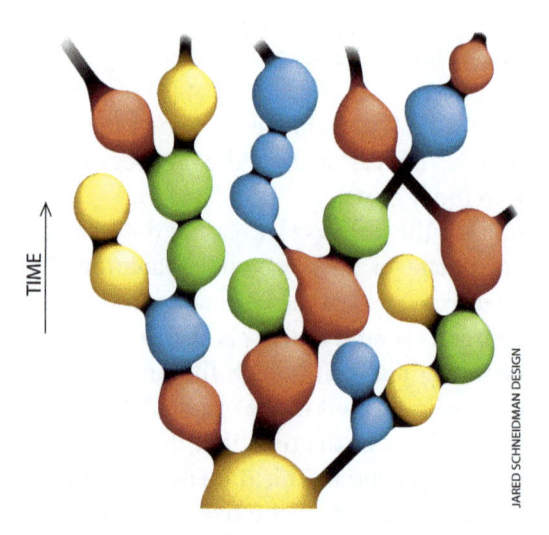

Fig. A3. Formation of mini-universes in chaotic inflation (*bubble theory*). Each universe has different physical laws, as shown by the different colors, and different space-time dimensionality. Credit: Jared Schneidman, *Scientific American*.

Each of these bubble universes will find itself at enormous distances from the others and will not be able to communicate with them.

What experimental predictions do the inflation theory have? We talked about it in Chapter 4. Here we mention them again.

- Inflation predicts that the universe is flat, $\Omega \approx 1$. If it started with Ω different than 1, inflation would drive it towards 1. CMB observations confirm this prediction. Furthermore, since the universe is flat, its energy is zero and therefore it could have formed from nothing, more precisely from quantum vacuum. To have more certainties on this and to open the Planck era door, a theory of quantum gravity is needed. String theory is moving away from this goal, but there is hope for loop quantum gravity.
- Like all quantum fields, inflation is subjected to continuous quantum fluctuations. There will be regions that come out of inflation first, while others emerge later. At the end of inflation, small differences in energy and density remain. These small density fluctuations grow due to gravity and are the seeds from which the structures were formed. Inflation provides a spectrum of perturbations with characteristics that are in accordance with what has been observed in the CMB map.
- As already mentioned, inflation involves the production of primordial gravitational waves. These, despite a false alarm from BICEP2 in 2014, have never been observed.
- Inflation entails a homogeneous and isotropic universe like the one observed.

Inflation explains several things, but leaves many things unexplained. For example, nothing is known about the field that originated it, the *inflaton*, which someone proposed to be the Higgs field.

In order for inflation to explain all just we described, it is necessary that the conditions that produced it were finely regulated (fine-tuned), a problem similar to that present in the hierarchy problem (see Chapter 6). According to Paul Steinhardt, one of the fathers of the theory, if the initial conditions are chosen randomly, it is more likely to obtain a "bad" inflation, that is, an inflation not capable to lead to the current universe. To determine with certainty that inflation happened, it is necessary to reveal the "first type" of gravitational waves produced by the "stretching" of the space due to the inflationary phase and the "second type" gravitational waves produced by the reheating phase. This would be the test, the "smoking gun," that the inflationary paradigm is correct. Those of the first type would have left marks on the CMB, but as already said, they have not been detected. Those of the second type with much greater amplitude have a frequency spectrum that exceeds the sensitivity threshold of the current instruments.

APPENDIX B: FORMATION OF STRUCTURES

In Chapter 2 we discussed the evolution of the universe from the Big Bang to the formation of the first stars. From the ashes of the primordial stars, a new generation of stars originated, the first galaxies and the first quasars. With telescopes, it has been possible to observe quasars and galaxies about a billion years after the Big Bang, and the merging of a quasar and a companion galaxy which happened about 12.8 billion years ago was recently observed. For the first time it was also possible to distinguish gas and stellar component of such a distant object, despite the presence of the dazzling quasar. How did these structures, and all the others, that make up the universe we observe form? They originated from quantum fluctuations, generated at the time of inflation, transformed by expansion into fluctuations in the density of matter and energy.

Another point of view is that advanced in 1977 by Hawking and Gibbons. The two physicists showed that in a universe with a *cosmological horizon*, exponentially expanding, as occurs in the cosmic inflation phase, thermal radiation must be emitted with a temperature linked to the rate of cosmological expansion. This radiation is connected to local variations in temperature. It can be concluded that primordial temperature fluctuations exist in the presence of accelerated expansion.

Quantum fluctuations, which are a sort of microscopic inhomogeneities, were amplified by the expansion in the period of inflation, generating the inhomogeneities of matter and temperature that are observed on the CMB. These inhomogeneities are density variations in the medium that make up the seeds from which the cosmic structures originated.

The physics of the formation of cosmic structures is complex and requires the use of numerical simulations for the study of the final stages. Simplifying:

- Quantum fluctuations are transformed into small density fluctuations and these grow during the era of radiation. The mass that makes up the universe is basically constituted of dark matter and in smaller quantities of baryonic matter. Of the two components, only the fluctuations of dark matter grow (i.e., the density increases) in almost all eras, while the perturbations of baryonic matter (electrons and protons) cannot grow until recombination because the baryons and radiation interact, forming acoustic waves. Thus, to have an understanding of how galaxies are formed, we first need to know how dark matter clusters. Indeed galaxies are immersed in huge regions made of dark matter, dubbed *halos*, which are formed, due to gravitational instability of the fluctuations of dark matter.
- The first halos that are formed are very small, and merge, forming larger and larger halos until they reach masses of the order of 10^{12} solar masses, which "help" baryonic matter to form galaxies like ours, or masses up to 10^{15} solar masses, that will form galaxy clusters.
- Initially there are invisible halos of dark matter that attract the baryons to fall inside them, cool and concentrate in the center, forming the visible component of the galaxies. Hence, all structures are formed by the gravitational collapse of small inhomogeneities in the material distribution, but not all of these inhomogeneities are capable of collapsing. Only those with mass, M, greater than a certain *Jeans mass*, collapse ($M > M_{Jeans} = 9/2 c_s^3/\sqrt{(\pi G \rho)} = (5kT/G\mu m_H)^{3/2}[3/(4\pi\rho)]^{1/2}$)[1]. The Jeans mass depends on the speed of sound cube (or square root of the temperature cube T) and on the square root of the density.

---

[1] Here c_s is the sound speed, ρ the density, T the temperature, μ the molecular weight, and m_H the hydrogen atom mass.

Appendix B: Formation of Structures

This mass is obtained when the gravity, which tends to make collapse the structure, and the internal pressure of the gas, which opposes the compression, are equal. If the mass of the cloud, M, is greater than Jeans mass, M_{Jeans}, it will collapse, while if it is lower, the gas under the combined action of the two forces will start to oscillate, forming density waves.[2] The Jeans theory is valid for a static system. In 1946 the Russian physicist Lifshitz extended this theory to the case of an expanding system, showing that the growth of inhomogeneities in the universe is slower than in a static system. In the case of star formation, the Jeans mass has values of about 1000 solar masses and the collapse occurs much more quickly than in the case of perturbations that form galaxies. This explains why the stars are not born in a solitary way, but in groups of thousands. To understand the formation of structures, it is necessary to know how the Jeans mass changes in the various cosmological epochs.

It is necessary to know what are the masses that can collapse and how the collapse grows over time. Both these factors depend on the cosmological era.

In Fig. B1 (left panel) we show how the Jeans mass of the baryons varies with time. Starting from a time corresponding to a value of the expansion parameter[3] $a \approx 10^{-10}$ (about 0.3 seconds), a little before the decoupling of neutrinos and the annihilation of positrons and electrons, up to $a \approx 10^{-6}$ (about a year), the Jeans mass of the baryons grows and is less than the typical mass of a galaxy, $\sim 10^{12} M_\odot$. Because $M \approx 10^{12} M_\odot > M_{Jeans}$, both for baryons and dark matter, the system is unstable and slowly collapses to form structures. In this phase, the perturbations of matter density, radiation, and dark matter grow together, as shown in Fig. B1 (right panel). Between $a \approx 10^{-6}$ and $a \approx 3 \times 10^{-4}$ (about 47,000 years), called the *era of equality*, the Jeans mass is greater than that of the galaxy ($M < M_{Jeans}$) and therefore the

---

[2] The oscillations have a wavelength proportional to the speed of sound in the medium and inversely proportional to the square root of the density.

[3] Recall that the expansion parameter, a, can be considered as the average distance between galaxies.

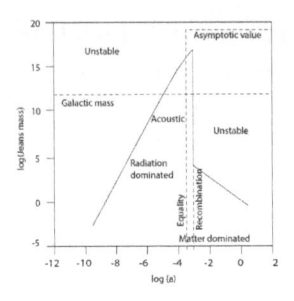

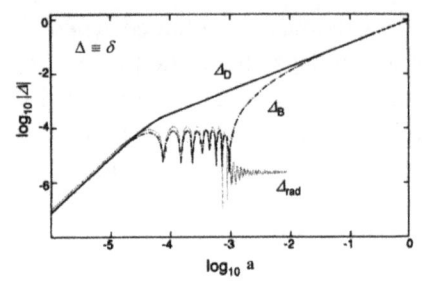

Fig. B1. Evolution of the Jeans mass. Up to about $a \sim 10^{-6}$ (about a year), an object such as the Milky Way with mass $M = 10^{12} M_\odot$ is unstable, that is, prone to collapse and form a structure, because $M > M_{Jeans}$, where M_{Jeans} is represented on the vertical axis. From $a \approx 10^{-6}$ to $a \approx 3 \times 10^{-4}$, this mass no longer grows because $M < M_{Jeans}$. After recombination, $a > 3 \times 10^{-4}$, we again have $M > M_{Jeans}$. The perturbation collapses and a new structure is formed. Right: Evolution of the fractional density $\delta = (\rho - \rho_c)/\rho_c$ of dark matter, baryonic matter, and radiation. Credit: E. Battaner and E. Florido (the rotation curves of spiral galaxies and its cosmological implications). Right: The evolution of δ at different epochs.

perturbations in matter and radiation cannot collapse to form structures.

Between $a \approx 3 \times 10^{-4}$ up to recombination (380,000 years), the photon–baryon interaction produces only oscillations, and there is no baryon structure growth. Instead, dark matter perturbations continue to grow. This perturbation behavior is shown in Fig. B1 (right panel), where oscillations and the continuous growth of dark matter are clearly shown.

After recombination the Jeans mass of the baryons falls to values smaller than those of the mass of a galaxy, $M \approx 10^{12} M_\odot$ much larger than Jeans mass, and the density perturbations of the baryonic matter begin to grow again, as shown in Fig. B1 (right panel), until it reaches the perturbations of dark matter, then collapsing to form structures. Since the dark matter density fluctuations have never stopped their growth, they have a greater density contrast and have already collapsed and formed halos, which act as holes (potential wells) in which ordinary matter falls, giving rise to structures, such as visible galaxies.

Appendix B: Formation of Structures

Historically, the first model of galaxy formation, due to Eggen, Lynden-Bell, and Sandage, assumed that the galaxies had formed from a *monolithic collapse* of a large gas cloud. This model fails to explain several of the characteristics of the galaxies. Searle and Zinn proposed a model in which galaxies are formed in an opposite process, a *hierarchical model*, starting from the coalescence of smaller progenitors, having masses of a few million suns, to form larger objects. The result of this process is the formation of a disk structure inside a halo of dark matter. In the hierarchical model, elliptical galaxies are formed by the merging of spiral galaxies. Among the various evidences in favor of the hierarchical model is the Hubble Deep Field (HDF), shown in section 2.6.

APPENDIX C: HIGGS MECHANISM IN MORE DETAIL

In section 6.7 we saw that for an electron with spin ½, we can have two *spin states*: ½ and –½. For massive particles with spin 1, like bosons W and Z, we could find the values –1, 0, 1. For a particle without mass (photon) we would find –1 and +1.

Spin states, polarizations, or *degrees of freedom* represent the ways in which a field can vibrate. At the same time, each polarization state can be considered as a different particle. In this sense, two particles "integrate" in an electron, and three particles in a W boson.

In summary, a field with spin 0, which we know to be a scalar field, has only one degree of freedom. A field with spin 1 without mass has two degrees of freedom, and as a consequence it can vibrate in a direction perpendicular to that of propagation, that is, up-down, right-left, as seen in the bottom box of Fig. C1.

A field with mass has three degrees of freedom, with two perpendicular excitations (1, –1) and one longitudinal (0), as shown in the top box of Fig. C1.

A massless particle like a photon moving at the speed of light cannot have a mode of vibration along the direction of motion because its speed cannot be larger or smaller than that of light. A particle such as a W or Z boson with mass, and therefore moving at a speed lower than that of light, does not have the problem of the photon and can "afford" to have a mode of vibration in the direction of motion.

In the primordial universe made up of massless particles that zipped at the speed of light, there were four mediators of the electroweak interaction: the *weak isospin fields*[4] W_1, W_2,

---

[4] Spin is associated with each elementary particle. It is also possible to associate a further conserved number, integer or half-integer, called *isospin*,

Appendix C: Higgs Mechanism in More Detail

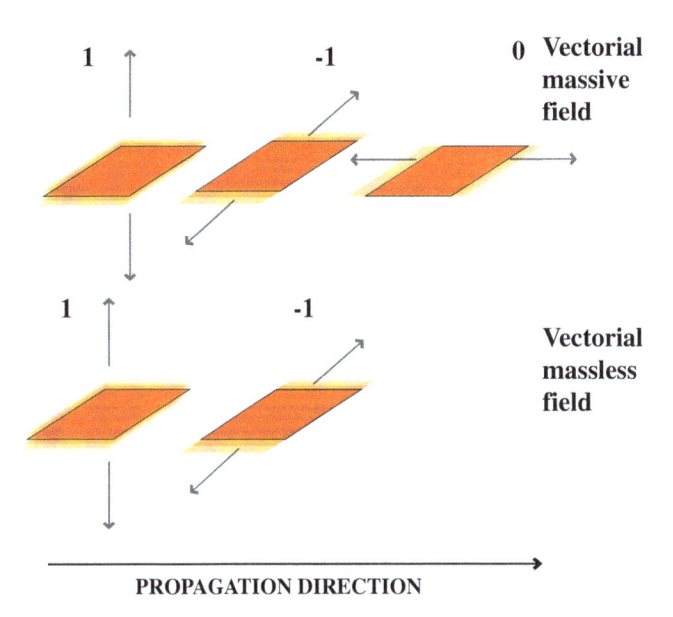

Fig. C1. Degrees of freedom of massive vector fields (top) and without mass (bottom). Credit: Modification of work by David Blanco, "Il bosone di Higgs," RBA Italia.

W_3, and the *weak hypercharge field*[5] B, all without mass. The problem is to create a world like ours in which there is the massless photon, γ, and the mediators W^+, W^-, Z^0, which have mass. As we said earlier, a massive vector boson needs three degrees of freedom. W_1, W_2, W_3 have two degrees of freedom and are devoid of the degree of longitudinal freedom. If we want them to receive mass, it is necessary to find three scalar particles which, by coupling with W_1, W_2, W_3, provide each with the degree of longitudinal freedom. The Higgs field has

---

which works like spin and regulates how the particle behaves in the weak interaction. The *weak isospin* has the same role as the electric charge in electromagnetism and that of color in the strong interaction.

It has three components, I_1, I_2, I_3, and is theoretically postulated to explain certain symmetries in decays.

[5] The hypercharge is the double of the difference between electric charge and isospin $Q = I_3^w + Y^w/2$, where *w* stands for weak.

just what is needed. It consists of four scalar particles: H_1, H_2, H_3, and H, which allow W_1, W_2, W_3 to become massive.

The first three, H_1, H_2, H_3, provide the degree of longitudinal freedom that W_1, W_2, W_3 lack in order to become massive. The last component of the Higgs field, H, does not couple with the photon (boson B), which remains massless, and we call it γ. H remains free and is the particle we call the *Higgs boson*. However, the complete mechanism implies not only that H_1, H_2, H_3 provide the longitudinal mode but also that the Higgs boson connects some components together. Before showing how this "connection" happens, we want to answer a question. How did the Goldstone bosons H_1, H_2, H_3 appear?

Higgs field played a fundamental role in the primordial universe. To better clarify what happened, we consider the evolution of the Higgs field (its equations of motion) as that of a ball that is initially at the top of the Mexican hat-shaped potential, shown in Fig. C2, and that then slides down. The height represents the potential energy, while the width at the base is the value (strength) of the field.

When the particle is on top of the Mexican hat, the potential energy is maximum, the field and mass of the particles is zero. The maximum position is symmetrical for

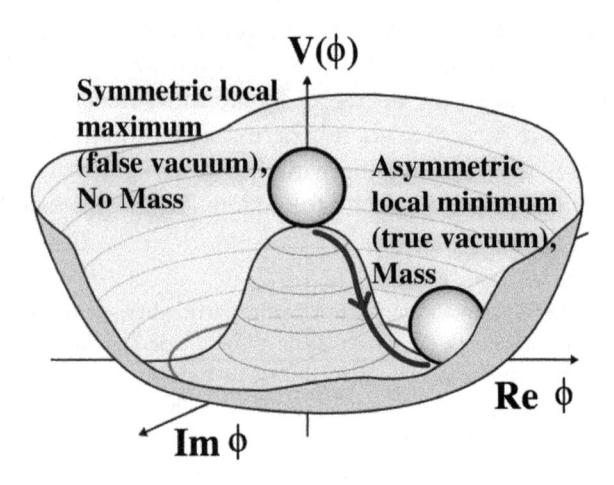

Fig. C2. Evolution of the Higgs field.

Appendix C: Higgs Mechanism in More Detail

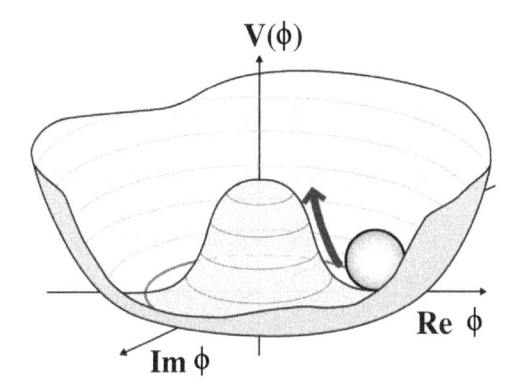

Fig. C3. Oscillations in the radial direction generating the massive Higgs boson, h. Credit: Modification of work by Flip Tanedo, Quantum Diaries.

rotation around the vertical axis passing through it, but it is very unstable and any quantum fluctuation can cause it to roll down from the hill to the minimum of the potential energy. When the ball, that is, the field, moves away from its rest position, the potential energy begins to transform into kinetic energy due to movement. When this happens, the field slides down to a random point on the Mexican hat base. The new position is no longer symmetrical, like the initial one. There was a *spontaneous symmetry breaking*.[6]

Before reaching its resting position, the ball swings up and down around the minimum and then stops in the minimum. The *up-and-down oscillations are the Higgs boson (H) (massive)* (Fig. C3).

When the Higgs H reaches the minimum, it acquires an average value in vacuum, equal to 246 GeV. When this happens, the Higgs field "is turned on" in all space and begins to do its "work," filling the whole space with its particles. Furthermore, when the field reaches the minimum, it can move on the whole circumference of the Mexican hat base (the green line), where

---

[6] The concept of spontaneous symmetry breaking is linked to the *phase transition* concept. As in a transition between liquid and solid state, symmetry breaking takes us from a perfectly symmetrical and massless world, at high temperature, to an asymmetrical and massive one at low temperature.

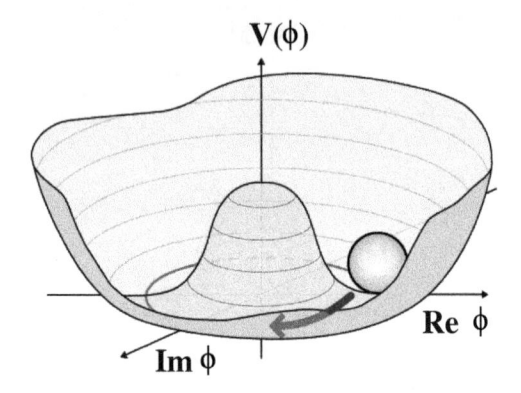

Fig. C4. Oscillations in the flat directions generating the massless Goldstone bosons H^+, H^-, and H^0. Credit: modification of work by Flip Tanedo, Quantum Diaries.

the value of the field is 246 GeV (the quantum mechanical ground state is called the *vacuum expectation value*), despite the fact that the field is in its rest position.

The Higgs has three flat directions corresponding to the Goldstone bosons (H^+, H^-, and H^0) (Fig. C4).

The up-and-down oscillations are the Higgs boson (H) (massive), while those along the base circumference are the massless Goldstone bosons, which will then be absorbed by the W_1, W_2, and W_3 bosons to form the gauge bosons W^+, W^-, Z^0.

In general its vibrations are coupled to the other fields, with a consequent transfer of energy from the Higgs field to the other fields, and particles get mass.

In addition to providing longitudinal modes, the Higgs makes a sort of connection between the bosons W_1, W_2, and W_3 and B, in a similar way to what it does with fermions. So, before describing the mechanism, I would like to remind you how the Higgs field supplies mass to fermions, for example electrons. In a simplified way, in Chapter 6, we said that the Higgs field behaves like a viscous substance for those particles. The particles entangled in the Higgs field move more slowly as if they were more massive. In slightly more technical terms, the description of the fields and particles in the standard model is carried out with a function called Lagrangian. The coupling of the Higgs with fermions gives an interaction term $-\lambda\psi^* H\psi$, which takes the name of *Yukawa interaction*, having a shape similar to that proposed by Yukawa in the 1930s. λ is the *Yukawa coupling constant*, H the Higgs field, ψ the fermion field, and ψ^* its *complex conjugate*. The mass of the fermion is proportional to the Yukawa coupling: the stronger the coupling force, the larger the mass of the fermion. When the spontaneous symmetry breaking occurs, the Higgs field goes down from the top of the hill (false vacuum) to the valley (true vacuum). When this happens, the Higgs field assumes its *vacuum expectation value* (an average energy value in the vacuum) of 246 GeV and the term $-\lambda\psi^* H\psi$ turns in $m\psi^*\psi$, where the mass m of the fermion is proportional to the value in the vacuum of the Higgs ($v = 246$ GeV) or $m = \lambda v/\sqrt{2}$.

Simplifying, the Higgs field connects the non-massive left-handed (see Appendix D) electron with the non-massive right-handed (see Appendix D) one to form the massive electron.

Let's go back to gauge bosons. Before supplying the longitudinal degrees of freedom to W_1, W_2, and W_3, making them become massive, the Higgs field connects them, as it does in the case of electrons. The complete scheme of the action of the Higgs is shown in Fig. C5.

The electromagnetic field is generated by the fields B and W_3. In this case, the Higgs field is limited to "intertwining" B and W_3, but does not provide the longitudinal component. As a result, one of the Higgs field bosons, H, remains free and is what we call the *Higgs boson*. The final result is a vector field with two components which therefore has no mass and moves at the speed of light: the photon, γ.

The photon is obtained by the superposition $\gamma = \sin\Theta_w W_3 + \cos\Theta_w B$, where Θ_w is called the *Weinberg angle*.

To generate the Z^0 field, the Higgs field intertwines B and W_3 and in addition provides the longitudinal component $H_1(H^0)$ (Fig. C5) to the new vector field, which has three components and is therefore massive. It turns out that $Z^0 = \cos\Theta_w W_3 - \sin\Theta_w B$. This relationship shows how the observable states (Z^0) have mass, but the underlying fields W_3 and B have no mass, respecting the gauge symmetry (invariance).

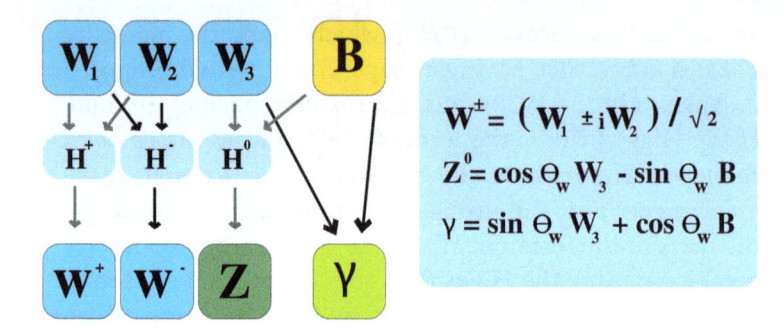

Fig. C5. The quanta of the Higgs field H_1, H_2, H_3, and H interact with the fields B, W_1, W_2, W_3, to generate the electromagnetic W^+, W^-, and Z^0 fields.

$$W^\pm = (W_1 \pm iW_2)/\sqrt{2}$$
$$Z^0 = \cos\theta_w W_3 - \sin\theta_w B$$
$$\gamma = \sin\theta_w W_3 + \cos\theta_w B$$

To generate W^+, the Higgs field intertwines W_1 and W_2 and provides the longitudinal component H_2 (H^+) (Fig. C5). The new field is the superposition of the three states W_1, W_2, and H_2.

To generate W^-, the Higgs field intertwines W_1 and W_2 and provides the longitudinal component H_3 (H^-) (Fig. C5). The new field is the superposition of the three states W_1, W_2, and H_3.

APPENDIX D: THE MASS OF NEUTRINOS

Neutrinos have no mass in the standard model, but their experimentally observed property of transforming into each other is possible only if they have a mass, although very small. The masses of elementary particles such as quarks and leptons are obtained by interacting with the Higgs field that couples the right- and left-handed versions of the particles. Unfortunately, neutrinos exist only in the left-handed form and therefore have no mass in the standard model.[7]

Furthermore, since the Higgs mechanism gives mass only to *Dirac particles*,[8] in the standard model the massive neutrinos, if they existed, should be only Dirac particles.

What does right-handed and left-handed mean? Let's consider a particle that moves and at the same time rotates around its axis, that is, it has spin. The sign of the projection of the spin vector, S, on the moment vector, p (product of mass by speed) is called *helicity*. If spin and speed have the same direction (Fig. D1 left side) the particle has a right-handed helicity, if they have opposite direction as in the right side of Fig. D1, the particle has left-handed helicity. To be precise, we would like to remind that helicity is independent of the reference system, that is, it does not transform from left to right, and vice versa, only for massless particles. For massive

Fig. D1. Helicity of a particle.

---

[7] Having no right-handed component, the Higgs mechanism cannot intertwine them as it does in electrons, providing them with mass.

[8] In case particle and antiparticle are the same object, we speak of Majorana particles.

particles, if we move faster than the particle, the helicity changes and for consistency we have to replace the concept of helicity with the more abstract one of *chirality*.[9]

The neutrino mass problem can be solved by assuming that they are antiparticles of themselves, that is, they are Majorana neutrinos, which could possess a so-called Majorana mass. However, even admitting that neutrinos are Majorana particles, it is necessary to explain why their masses are so small.

A proposed solution is linked to the existence of right-handed neutrinos, such as the so-called sterile neutrino, already mentioned. These neutrinos could interact with the Higgs field as other fermions do, without being subject to weak interaction. If these neutrinos had a very large mass, the small mass of the usual neutrinos would be explainable, since according to the seesaw mechanism, the mass of the usual neutrinos is inversely proportional to that of the sterile neutrino.

More precisely, the mass of the usual neutrinos is given by the ratio between the square of the electroweak energy scale, ~246 GeV, and the GUT energy scale, 10^{16} GeV, a ratio that is worth about a hundredth of eV (10^{-2} eV), exactly the expected value for the neutrino masses. This mechanism is based on the existence of new physics at an energy scale 10,000–100,000 billion times larger than that of the electroweak scale (100 GeV) (i.e., the scale at which the Higgs boson was found), which can be the scale of the Grand Unification Theory, 10^{15}–10^{16} GeV.

---

[9] This term comes from the Greek χείρ, "cheir" (hand). An image reflected by a mirror cannot be superimposed on itself, like the case of a hand. More precisely, a system is chiral if applying a parity transformation — that is, inverting the spatial coordinates (from (x, y, z) to $(-x, -y, -z)$), the system does not overlap with itself and changes chirality, from right to left, and vice versa. Until 1956, the universe was thought to be invariant by operating parity symmetries, or, in other words, there were no processes of a nature that could distinguish between its real image and the reflected one. In 1956 the Chinese physicist Wu dissipated Lee and Yang's doubts that nature does not always respect the principle of parity, showing that parity is not preserved in weak interactions.

BIBLIOGRAPHY

Aczel, A. D. (1999). God's Equation: Einstein, Relativity and the Expanding Universe.

Alonso, J. M. (1989). *Introducion al principio antropico* (Ediciones Encuentro, S.S., Madrid).

Barbour, J. (2005). The End of Time: The Next Revolution in Physics, Oxford University Press.

Bertone, G. (2013). Behind the Scenes of the Universe: From the Higgs to Dark Matter, Oxford.

Carrol, S. (2012) The Particle at the End of the Universe: How the Hunt for the Higgs Boson Leads Us to the Edge of a New World, Dutton.

Carrol, S. (2010). From Ethernity to here; The quest for the Ultimate theory of time, Oneworld.

Casas, A. (2016). *La materia oscura* (Dark matter) (RBA Italia, Milano).

Davies, P. (1993). The Mind of God: The Scientific Basis for a Rational World.

Davies, P. (2008). *Superstrings: A Theory of Everything?* (Cambridge University Press).

Deutsch, D. (1997). The fabric of reality, Penguin Books.

Evans, R. The Cosmic Microwave Background: How It Changed Our Understanding of the Universe, Springer.

Ferreira, P. (2014).The Perfect Theory: A Century of Geniuses and the Battle over General Relativity, Mariner Book.

Feynman, R. P. (1986). Qed: The Strange Theory of Light and Matter (Princeton Science Library).

Gregg, B. (2019), Dark matter and dark energy, the hidden 96% of the universe, Hotscience.

Guth, A. (1998), The Inflationary Universe.

Gardner, M. (2003). *Multiverses and Blackberries* (W. W. Norton, New York).

Green, B. (2011). The Hidden Reality: Parallel Universes and the Deep. Laws of the Cosmos, Penguin Books.

Greene, B. (2000). The Elegant Universe: Superstrings, Hidden Dimensions, and the Quest for the Ultimate Theory, Vintage Books.

Hawking, S. A Brief History of Time: From Big Bang to Black Holes

Hawking, S., *Mlodinow,* L., The Grand Design.

Hoddeson, L., *et al.* (1997). The Rise of the Standard Model: A History of Particle Physics from 1964 to 1979.

Hogan, C. J. (1998). *The Little Book of the Big Bang: A Cosmic Primer* (Springer-Verlag, New York).

Impey, C. (2010). How It Ends: From You to the Universe, Norton & Company.

Kaku, M. (2006). Parallel Worlds: A Journey Through Creation, Higher Dimensions, And the Future of the Cosmos.

Kragh, H. (2013). *Conceptions of Cosmos: From Myths to the Accelerating Universe — A History of Cosmology* (Oxford University Press, Oxford).

Krauss, L. M. (2012). A Universe from Nothing: Why There Is Something Rather than Nothing, Free Press.

Luminet, J.-P. (2006). *L'invenzione del big bang: storia dell'origine dell'universo* (Dedalo, Bari, Italy).

Mitton, S. and Ostriker, J. (2015). *Heart of Darkness: Unraveling the Mysteries of the Invisible Universe* (Princeton University Press).

Ne'eman, Y. and Kirsh, Y. (1988). The Particle Hunters.

Neuenschwander, D. E. (2010). *Emmy Noether's Wonderful Theorem* (Johns Hopkins University Press, Baltimore).

Pagels, H. R. (1984). The Cosmic Code: Quantum Physics as the Language of Nature, Dover Publications.

Parsons, P. (2018). The Beginning and the End of Everything: From the Big Bang to the End of the Universe.

Penrose, R. The Road to Reality: A Complete Guide to the Laws of the Universe.

Profumo, S. (2017). An Introduction to Particle Dark Matter, World Scientific.

Randall, L. (2012). Knocking on Heaven's Door: How Physics and Scientific Thinking Illuminate the Universe and the Modern World, Harper Collins.
Randall, R. (2016). Dark Matter and the Dinosaurs: The Astounding Interconnectedness of the Universe.
Rees, M. (2001). Just Six Numbers: The Deep Forces That Shape The Universe.
Rovelli, C. (2017). Reality is not like it seems, Penguin Books.
Susskind, L. The Cosmic Landscape: String Theory and the Illusion of Intelligent Design.
Tegmark, M. (2015). Our Mathematical Universe: My Quest for the Ultimate Nature of Reality.
Thorne, K. S. (1994). Black Holes & Time Warps: Einstein's Outrageous Legacy, Norton & Company.
Thuan, T. X. (1995). The Secret melody, and men created the Universe, Oxford, University Press.
Vilenkin, A. Many Worlds in One: The Search for Other Universes
Weinberg, S. (1980). The First Three Minutes: A Modern View Of The Origin Of The Universe.
Weinberg, S. (2010). *Dreams of a Final Theory: The Scientist's Search for the Ultimate Laws of Nature* (Random House, New York).
Weyl, H. (1952), Symmetry, Princeton Science Library.

INDEX

absolute magnitude, 19
absorption spectrum, 13
abundance of light elements, 241
acoustic peaks, 92
adiabaticity, 96
AMANDA, 190
AMS-02, 189
annihilation cross section, 162
anthropic principle, 212
anthropos, vii
anti-de Sitter space, 170
anti-leptons, 38
antiparticle pairs, 38
antiparticles, 36
apparent luminosity, 19
arcs, 83
argon, 181
astronomical unit, 5
Astronomy with a Neutrino Telescope and Abyss Environmental Research (ANTARES), 190
asymmetric dark matter, 172
asymptotic freedom, 146
atom, 115

atomist theory, 115
axion, 36, 112
axion dark matter experiment (ADMX), 185

background imaging of cosmic extragalactic polarization (BICEP), 97
background noise, 178
background radiation sound spectrum, 94
backreaction, 222
Baikal Deep Underwater Neutrino Telescope (BDUNT), 190
balloon observations of millimetric extragalactic radiation and geophysics (BOOMERANG), 48, 205
baryon acoustic oscillations (BAO), 208
baryonic matter, 6
baryonic number, 149
baryon oscillation spectroscopic survey (BOSS), 209

baryons, 32, 38, 147
beta decay, 135
beryllium, 41
Big Bang, vii, 7, 22, 23, 26, 27, 45, 48, 53, 87
Big Bang cosmology, 36
Big Bang model, 7
Big Bounce, 26, 234
Big Crunch, xi, 26, 196
Big Rip, 219
Big Slurp, 229
black body, 47
black dwarf, 105
black hole, 7, 65, 104, 106
blueshift, 14
B modes, 238
Bose-Einstein statistic, 146
bosonic fields, 120
bosons, 32
bottom-up models, 113
brane, 27
brane multiverse, 217
breaking scale of supersymmetry, 163
Brown dwarfs, 104
Brownian motion, 116
brown, red, white dwarfs, 100
brown sub-dwarfs, 104
bubble theory, 239, 246
bulge, 59
bulk, 170
bullet cluster, 84, 85

Calabi–Yau spaces, 166
Calorimetric Electron Telescope (CALET), 189
CANGAROO-III, 187
carbon cycle, 56
Casimir effect, 123

CDMSI, II, 180
Cepheids, 18, 19, 196
CERN Solar Axis Telescope (CAST), 185
Chameleon screening, 220
Chandrasekhar limit, 105, 197
Cherenkov radiation, 187
chirality, 262
chromosphere, 101
clockwork universe, 1
closed universe, 6
clusters, 73
clusters of galaxies, 61
CMB anisotropies, 48
CMB map, 87
COBE satellite, 47
COGENT, 180
cold dark matter, 36
cold dark matter model, 112
color, 145
Coma, 73
Coma Cluster, 67, 73, 99
confinement, 146
conformal cyclic cosmology, 238
connection field, 129
conservation law, 128
conservation of energy, 128
conservation of the momentum, 128
continuous symmetry, 126
continuum, 115
CORE, 97
corona, 101
cosmic background of neutrinos, 38
cosmic microwave background radiation (CMB), 22, 23, 36, 42, 45, 87, 97
cosmic rays, 67

Index

cosmic shear, 209
cosmic web, 61
cosmological constant, 3
cosmological constant problem, 205
cosmological horizon, 26, 249
couple-instability supernovae, 56
CP symmetry, 39
CRESST, 180
critical density, 4, 6
cryogenic experiments, 180
Cryogenic Dark Matter Search (CDMS), 113, 180
Cherenkov Telescope Array (CTA), 187
curvature, 6
curvature of space, 3
cusp–core problem, 173
cyanogen, 45
cyclic universe, 26

DAMA/LIBRA experiment, 178
dark energy, 94, 96, 202
dark matter, 65, 67, 74, 87
Dark Matter Explorer (DAMPE), 189
dark matter halos, 58
dark photon, 171, 172
DARKSIDE, 181
DARKSIDE-20k, 181
DARWIN, 182
D-branes, 169
decay of the proton, 159
decoupling, 43
deflection of light, 80
degenerate electrons, 103, 104
degenerate state, 104

degrees of freedom, 138
density fluctuations, 49, 96
density parameter, 6
detection channels, 178
deuterium, 54
dipolar anisotropy, 46
dipolar repulsor, 46
Dirac particles, 261
direct detection, 175
discrete symmetry, 126
disk-halo conspiracy, 72
DMICE experiment, 180
doppler effect, 11, 14, 20, 70
dwarf galaxies, 188
dwarf spheroidal, 59
dwarf stars, 100

EDELWEISS, 180
EGRET, 192
Einstein condensate, 147
Einstein cross effect, 81
Einstein ring, 81
ekpyrotic model, 27
ekpyrotic universe, 37, 237
electric field, 10
electromagnetic field, 121
electromagnetic force, 33, 151
electromagnetic spectrum, 10
electromagnetic wave, 10
electron neutrinos, 110
electrons, 32, 33, 38
electronvolt, 5
electroweak era, 36
electroweak interaction, 139
electroweak symmetry breaking, 37
electroweak theory, 117
elliptical galaxies, 59
emission spectrum, 13

end of time, 228
entropy, 234
equation of state, 219
equation of state parameter, 219
era of equality, 251
era of nucleosynthesis, 40
era of radiation, 42
era of recombination/
    decoupling, 42
EROS, 108
escape velocity, 7
eternal chaotic inflation, 215
eternal inflation, 216, 246
eternal return, 1
Euclidean geometry, 6
Euclid mission, 222
event horizon, 77, 106
exclusion curves, 183
expanding universe, 20
expansion parameter, 4
extra dimensions, 163
extrasolar planets, 102, 213
eXtreme Deep Field (XDF), 58

false vacuum, 122
FERMI bubbles, 193
Fermi–Dirac statistic, 146
Fermi energy, 147
FERMI satellite, 186
fermionic fields, 120
fermions, 32, 33, 146
field, 34
fifth force, 220
filaments, 61
fine-tuning problem, 157
first peak, 93
flat geometry, 30
flux, 19
fossil quantity, 162

fossil radiation, 17
freeze-out, 161
frequency, 10
Friedman's closed universe, 7
Friedman's solutions, 6
Friedman's Universe, 3
fundamental fields, 117
fundamental frequency, 89
fuzzy dark matter, 171, 172

gauge bosons, 133, 147
gauge field, 129
gauge symmetries, 129
gauge theory, 136
gauge transformation, 126
gaussianity, 96
general relativity, 2, 32
geodesics, 77
global transformation, 126
gluino, 155
gluons, 32, 146
Goldilocks Zone, 103
Goldstone bosons, 139
Goldstone's theorem, 137
grand unification, 158
grand unification age, 34
grand unification phase
    transition, 35
grand unification theory
    (GUT), 35, 158, 159
gravitational field, 84
gravitational field equations, 75
gravitational force, 33, 151
gravitational lens effect, 67
gravitational lens equation, 84
gravitational potential, 120
gravitational singularity, 7
gravitational waves, 227
graviton, 112, 155, 159

great attractor, 46
Great Magellanic Cloud, 226
great unification phase transition, 35
greenhouse effect, 226
Gunn–Peterson effect, 51

habitable zone, 103
hadronic era, 38, 39
hadrons, 32, 38, 147
halos, 250
harmonics, 89
Hartle–Hawking state, 27
Hawking radiation, 111, 167, 227
HDM model, 112
heavy leptons, 155
Heisenberg uncertainty principle, 121
helicity, 261
helium-4 nucleus, 39
hidden or dark sector, 172
hidden symmetry, 128
hierarchical model, 59, 253
hierarchy problem, 151, 156
Higgs boson, 33, 117, 151
Higgs field, 36, 123
higgsino, 155
Higgs mechanism, 137, 138
Higgs portal, 172
High-Altitude Water Cherenkov Observatory (HAWC), 187
High Energy Stereoscopic System (HESS), 187
HI line, 50
holographic multiverse, 217
holographic principle, 217
homogeneous, isotropic and non-static universe, 4

homo sapiens, vii
horror vacui, 115
hot dark matter model, 112
Hubble constant, 21, 25
Hubble Deep Field (HDF), 58
Hubble's law, 9, 15, 23, 24, 195
Hydra–Centaurus supercluster, 46
hyperbolic geometry, 6
hyperbolic universe, 7

imaginary time, 27
indirect detection, 176, 185
inflation, 36, 37, 88, 96
inflation theory, 30, 35
inflaton, 35, 96, 215, 243
initial spectrum of perturbations, 36
integral, 193
interaction field, 129
interactions, 33
internal symmetries, 128
invariant, 126
ionization, 12
ionization channel, 180
ionization detection, 177
irregular galaxies, 59
isospin, 254
isotope, 55

JAXA, 97
Jeans mass, 101, 250

Kaluza–Klein (KK) tower, 169
Kaluza–Klein theory, 163
Kaluza–Klein tower of states, 171
Kant–Laplace's nebular hypothesis, 102
kaon, 145

Kelvin, 5
Keplerian fall, 70
kinetic quintessence, 219
KM3NET, 190

Lagrangian, 258
Lamb effect, 123
landscape multiverse, 217
large-scale structure, 61
latent heat, 35
Lemaître–Tolman–Bondi model, 222
lensing effect, 80
leptonic era, 38
leptons, 33, 38, 145, 149
large hadron collider (LHC), xi, 191
light-year, 5
LIGO/VIRGO collaboration, 109
lines of force, 119
LiteBIRD, 97
lithium, 104
loop quantum theory, 26
LUX, 182

μ meson, 144
MACHO dark matter, 100
magnetic field, 10
main sequence phase, 105
Major Atmospheric Gamma Imaging Cherenkov Telescope (MAGIC), 187
many-worlds interpretation of quantum mechanics, 215
massive astrophysical compact halo objects (MACHO), 100, 108
mesons, 33, 38, 147, 149

metastable state, 231
Mexican hat potential, 215
microlensing effect, 83
microwave background radiation, xi
minimal supersymmetric model (MSSM), 159
missing satellites problem, 173
modified gravity, 221
modified Newtonian dynamics (MOND) theory, 221
moment, 33
monolithic collapse, 253
M theory, 27, 169
multiverse, 37, 214
muon neutrinos, 110
muons, 32, 33

naked mass, 156
Nambu–Goldstone bosons, 137, 138
negative pressure, 200
neutralino, 159
neutrino, 32, 100, 136
neutrino background, 38
neutrino oscillations, 33, 112
neutrons, 32, 38
neutron star, 67, 101, 104
Newtonian mechanics, 2, 6, 75
Newton's Principia, 119
non-conservation of the baryon number, 159
nucleosynthesis, 41, 50

observable universe, 26
Olbers paradox, 2
Oort cloud, 198
Oort limit, 67
open universe, 6

optical gravitational lensing experiment (OGLE), 108
orbitals, 131
ordinary dark matter, 99

pancakes, 112
parallel universes, 218
parity symmetry, 136
parity transformation, 136
Parsec, 5
particles, 36
Pauli exclusion principle, 104, 146
PAVLAS collaboration, 185
payload for antimatter matter exploration and light-nuclei astrophysics (PAMELA), 189
peak (harmonics), 90
perfect cosmological principle, 7
periodic world, 5
perturbation theory, 132, 156
phantom energy, 232
phantom energy model, 219
phase transition, 257
Phillips's supernovae, 199
phoenix universe, 37
phonons, 176
photino, 112, 155
photoelectric effect, 130
photomultiplier tubes, 182
photons, 29, 32, 34, 38
photosphere, 101
PIXIE, 97
PLANCK, 48, 49, 53, 68, 97
Planck epoch, 31
Planck era, 32, 35
Planck mass, 156
Planck mission, 25

Planck time, 30, 31, 32
planetary nebula, 105
planetesimals, 102
plasma of quarks, 37
p meson, 145
polarization modes, 96
polarization states, 138
polarized light, 185
positrons, 38
power of 10, 5
precession of the perihelion of Mercury, 64
Primakoff effect, 185
primeval atom, 7
primordial atom, 15, 17
primordial black holes, 109
primordial gravitational wave background, 36
primordial gravitational waves, 96
primordial nucleosynthesis, 109
primordial universe, 29
principia matematica, 1
principle, 34
principle of equivalence, 75
problem of coincidence, 212
problem of flatness, 243
problem of the horizon, 243
problem of topological defects, 243
proton–proton chain, 54
protons, 32, 38
protoplanetary disk, 102
protostar, 102
pulsars, 106

Q-balls, 171
quantum chromodynamics, 117, 146

quantum electrodynamics
(QED), 117, 124, 133, 134
quantum entanglement, 97
quantum field, 30, 34, 120
quantum field physics, 204
quantum field theories, 117
quantum fluctuations, 33, 88, 96
quantum gravity, 30, 32, 159
quantum jumps, 131
quantum loop theory, 228
quantum mechanics, 32
quantum multiverse, 217
quantum tunneling, 230
quantum uncertainty, 34
quantum vacuum, 30, 32
quark era, 38
quark–gluons plasma, 37
quark model, 145
quarks, 32
QUIJOTE, 97
quintessence, 218

radiative and a convective region, 100
radio band, 72
Randall–Sundrum theory, 170
recombination, 43
red giant, 103, 226
redshift, 9, 11, 14, 46
reductionism, 1
re-heating phase, 35, 245
renormalization, 133
rings, 81
rotation curve, 69
R-parity, 160

Saros cycle, ix
Sauter–Schwinger effect, 123, 124
scalar field, 34, 117, 244
scalar-type field, 219
scintillation, 178
scintillation of the sodium iodide crystals, 179
screening, 220
second peak, 93
seesaw mechanism, 150
selectron, 156
self-interacting dark matter, 172
Shapley Attractor, 46
Silk damping, 93
simulated multiverse, 217
slow rolling phase, 245
Snell–Descartes law, 74, 84
sneutrino, 156
snow line, 102
solar wind, 100
sound horizon, 209
sound spectrum, 88, 91, 92
special relativity, 133
spectral index, 96
speed of recession, 9
spin, 32
spin-independent scattering, 176
spin states, 138
spin-statistic theorem, 146
spiral galaxy, 59, 69
spontaneous breaking of symmetry, 137
spontaneous symmetry breaking, 126, 257
squark, 156
standard candles, 19, 196
standard model of elementary particles, xi
standard model of particles, 113, 116

standard ruler, 209
state parameter, 232
steady-state theory, 7
stereoscopic technique, 187
sterile neutrino, 150, 171
string theory, 26
strong, 83
strong lensing, 83
strong nuclear force, 33, 134, 151
substructures, 188
superCDMS, 180
superCDMS SNOLAB, 180
Super Cluster of Laniakea, 46
superclusters, 61
superconducting material, 176
super-force, 158
supergiants, 101
super-Kamiokande, 227
supernova, 67, 104
supernova cosmology project, 198
superstring theory, 164
supersymmetry (SUSY), xi, 155, 157
surface of last scattering, 48, 92
symmetry breaking, 127
symmetry of charge and parity, 39

2-sphere, 3
3-sphere, 2, 5
table of the elements, 116
tau, 32
tau meson, 145
tau neutrinos, 110
tauons, 33
theories of grand unification, 227

theories of modified gravity, 65
theory of everything, 164
theory of extra-large dimensions, 169
thermal death, 225
the standard model of particles, 117
the virtual particles, 121
third peak, 93
tomographic reconstruction, 211
too-big-to-fail problem, 173
top-down model, 112
towers of particles, 168
tritium, 55
true vacuum, 122
twilight of the gods,, 223
type Ia supernovae, 197

ultraviolet (UV) radiation, 56
uncertainty, 34
uncertainty principle, 33
unification of fundamental forces, 157
universal extra dimension (UED), 171
universe, 1

vacuum, 34
vacuum expectation value, 258
vacuum of the field, 122
Vainshtein screening, 220
vector field, 34
vectorial field, 117
Very Energetic Radiation Imaging Telescope Array System (VERITAs), 187
Virgo cluster, 61, 73
virial theorem, 67, 74

virtual gluons, 133
virtual particles, 34
voids, 61

warm dark matter, 96
wave equation, 131
wave function, 131
wavelength, 10, 29
wave-particle dualism, 11, 130
way of the octet, 145
weak, 83
weak hypercharge field, 255
weak isospin fields, 254
weak lensing effect, 83
weakly interacting massive particles (WIMPs), 160, 175
weak nuclear force, 33, 134, 151
white dwarf, 103

Wilkinson Microwave Anisotropy Probe (WMAP), 48, 97
WIMPs miracle, 162
Wimpzillas, 171, 172
wino, 155
wormholes, 31

X and Y bosons, 32
xenon, 182
XENON, 179, 182
XEXON1T, 182
X-ray band, 73

Yukawa coupling constant, 259
Yukawa's interaction, 141

zino, 155

CPSIA information can be obtained
at www.ICGtesting.com
Printed in the USA
BVHW050224130522
636929BV00003B/28

9 789811 252631